Chemists in a social and historical context

Chemists are real people, living in the real world

Written by Dorothy Warren
RSC School Teacher Fellow 1999–2000

ROYAL SOCIETY OF CHEMISTRY

Chemists in a social and historical context

Written by Dorothy Warren

Edited by Colin Osborne and Maria Pack

Designed by Imogen Bertin

Published and distributed by Royal Society of Chemistry

Printed by Royal Society of Chemistry

Copyright © Royal Society of Chemistry 2001

Registered charity No. 207980

For further information on other educational activities undertaken by the Royal Society of Chemistry write to:

Education Department
Royal Society of Chemistry
Burlington house
Piccadilly
London W1J 0BA

Information on other Royal Society of Chemistry activities can be found on its websites:
http://www.rsc.org
http://www.chemsoc.org
http://www.chemsoc.org/LearnNet contains resources for teachers and students from around the world.

ISBN 0–85404–380–2

British Library Cataloguing in Publication Data.

A catalogue for this book is available from the British Library.

RS•C

Foreword

This resource is designed to allow young people to see science as a social activity and to give examples of the nature of science from a historical perspective. The examples have been chosen to both reflect the serendipitous nature of some scientific discoveries and also the multi-cultural nature of society. It is hoped that a judicious choice of some of the examples will give students an appreciation of science as an exciting activity and provide an extra stimulus for them to engage in science themselves.

Professor Steven Ley CChem FRSC FRS
President, The Royal Society of Chemistry

RS•C

Acknowledgements

The production of this book was only made possible because of the advice and assistance of a large number of people. To the following, and everyone who has been involved with this project, including the members of the science staff and students in trial schools, both the author and the Royal Society of Chemistry express their gratitude.

General
Colin Osbome, Education Manager, Schools & Colleges, Royal Society of Chemistry
Maria Pack, Assistant Education Manager, Schools & Colleges, Royal Society of Chemistry
Members of the Royal Society of Chemistry Committee for Schools and Colleges.
Members of University of York Science Education Group.
Jill Bancroft, Special educational needs project officer, CIEC
Donald Stewart, Dundee College, Dundee
Richard Warren, Mathematics Department, Ampleforth Colledge, York
Bob Campbell, Department of Educational Studies, University of York
Professor David Waddington, Department of Chemistry, University of York.
Graham Wright, York Sugar Factory
The Patent Office, Harmsworth House, 13–15 Bouverie Street, London.
David Taylor, Corus Technology, Rotherham

Pictures
Nobel Foundation http://www.nobel.se
Crystallographica (Oxford Cryosystems)
The Schomburg Center, New York Public Library
Hagley Museum and Library
The Library and Information Centre, Royal Society of Chemistry.

Schools
Sandra Buchanan, Tobermory High School, Isle of Mull
Arthur Cheney, All Saints School, York.
Tim Gayler, Little Ilford School, London.
Louise Campbell, Greencroft School, Stanley.
Margaret Crilley, St Leonard's RC Comprehensive School, Durham.
John Davies, Hipperhoime & Lightcliffe High School, Halifax.
John Ediin, Wolverhampton Grammar School, Wolverhampton.
Greg McClarey, Blessed Edward Oldcome RC High School, Worcester
Joy Page, Clevedon Community School, Somerset
Lesley Stanbury, St Albans School, St Albans.
Lynne Tomes, Ysgol Llanilltud Fawr, Vale of Glamorgan

The Royal Society of Chemistry would like to extend its gratitude to the Department of Educational Studies at the University of York for providing office and laboratory accommodation for this Fellowship and the Head Teacher and Governors of Fulford Comprehensive School, York for seconding Dorothy Warren to the Society's Education Department.

Contents

RS•C

How to use this resource

At the start of the 21st century secondary education yet again underwent changes. These included the introduction of new curricula at all levels in England, Wales and Scotland and the Northern Ireland National Curriculum undergoing review. With more emphasis on cross curricula topics such as health, safety and risk, citizenship, education for sustainable development, key skills, literacy, numeracy and ICT, chemistry teachers must not only become more flexible and adaptable in their teaching approaches, but keep up to date with current scientific thinking. The major change to the science 11–16 curricula of England and Wales was the introduction of 'ideas and evidence in science', as part of Scientific Enquiry. This is similar to the 'developing informed attitudes' in the Scottish 5–14 Environmental studies, and is summarised in Figure 1.

In this series of resources, I have attempted to address the above challenges facing teachers, by providing:

■ A wide range of teaching and learning activities, linking many of the cross-curricular themes to chemistry. Using a range of learning styles is an important teaching strategy because it ensures that no students are disadvantaged by always using approaches that do not suit them.

■ Up-to-date background information for teachers on subjects such as global warming and Green Chemistry. In the world of climate change, air pollution and sustainable development resource material soon becomes dated as new data and scientific ideas emerge. To overcome this problem, the resources have been linked to relevant websites, making them only a click away from obtaining, for example, the latest UK ozone data or design of fuel cell.

■ Resources to enable ideas and evidence in science to be taught within normal chemistry or science lessons. There is a need to combine experimental work with alternative strategies, if some of the concerns shown in Figure 1, such as social or political factors, are to be taught. This can be done for example, by looking at the way in which scientists past and present have carried out their work and how external factors such a political climate, war and public opinion, have impinged on it.

■ Activities that will enhance student's investigative skills.

These activities are intended to make students think about how they carry out investigations and to encourage them to realise that science is not a black and white subject. The true nature of science is very creative, full of uncertainties and data interpretation can and does lead to controversy and sometimes public outcry. Some of the experiments and activities will be very familiar, but the context in which they are embedded provide opportunities for meeting other requirements of the curriculum. Other activities are original and will have to be tried out and carefully thought through before being used in the classroom. Student activities have been trialled in a wide range of schools and where appropriate, subsequently modified in response to the feedback received.

Dorothy Warren

RS•C

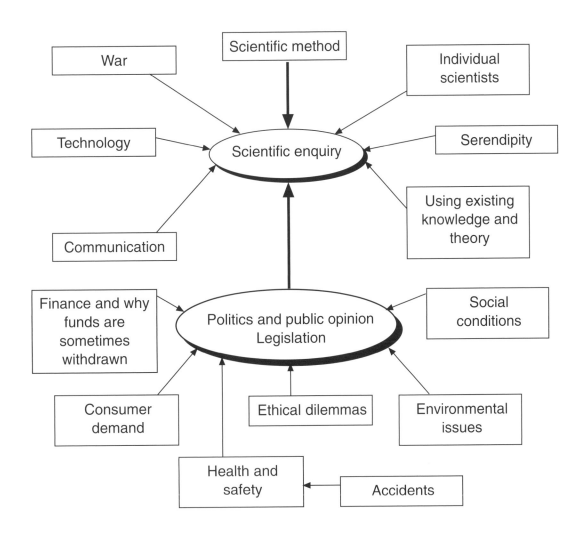

Figure 1 The factors influencing the nature of scientific ideas– scientific enquiry and the advancement of science

Maximising the potential use of this resource

It is hoped that this resource will be widely used in schools throughout the United Kingdom. However, as every teacher knows, difficulties can be experienced when using published material. No single worksheet can cater for the needs of every student in every class, let alone every student in every school. Therefore many teachers like to produce their own worksheets, tailored to meet the needs of their own students. It was not very surprising when feedback from trial schools requested differentiated worksheets to allow access to students of different abilities. In an attempt to address these issues and concerns, this publication allows the worksheet text and some diagrams to be modified. All the student worksheets can be downloaded in Word format, from the Internet via the LearnNet website, **http://www.chemsoc.org/networks/learnnet/ideas-evidence.htm** .This means that the teacher can take the basic concepts of the activity, and then adapt the worksheet to meet the needs of their own students. Towards the end of the teachers' notes for most activities there are some suggestions as to how the resource can be adapted to meet the needs of students of different abilities. There are also some examples of differentiated worksheets included in the resource.

RS•C

It is not envisaged that teachers will use every activity from each piece of work with an individual class, but rather pick and choose what is appropriate. For example some activities use high level concepts and are designed to stretch the most able student and should not be used unchanged with less able students eg **The atom detectives** and **What happens when things burn?**

Activities that involve researching for secondary information on the Internet contain hyperlinks to appropriate websites. To minimise the mechanical typing of the URLs and possible subsequent errors, the students can be given the worksheet in electronic form and asked to type in their answers. The websites are then only a click away.

Appropriate secondary information has been included in the teachers' notes for use in class when the Internet or ICT room is unavailable.

Unfortunately, from time to time website addresses do change. At the time of publication all the addresses were correct and the date that the site was last accessed is given in brackets. To minimise the frustration experienced when this happens, it is advisable to check the links before the lesson. If you find that a site has moved, please email both **LearnNet@rsc.org** and **education@rsc.org** giving full details so that the link can be updated on the worksheets on the web in the future.

Strategies for differentiated teaching

All students require differentiated teaching and it is not just an issue for those students with special educational needs. The following definition by Lewis[1] has been found to be quite useful.

'Differentiation is the process of adjusting teaching to meet the needs of individual students.'

Differentiation is a complex issue and is very hard to get right. It can be involved in every stage of the lesson ie during planning (differentiation by task), at the end of the activity (differentiation by outcome) and ongoing during the activity. Often teachers modify the activity during the lesson in response to feedback from the class. Differentiation does not only rely on appropriate curriculum material but is also concerned with maximizing learning. Student involvement and motivation effect the learning experience and should be considered and taken into account. It is therefore not surprising that differentiation is one of the areas of classroom teaching where teachers often feel under-confident. Most strategies for differentiated lessons are just applying good teaching practice eg varying the pace of the lesson, providing suitable resources and varying the amount and nature of teacher intervention and time.[2] Rather than just providing several examples of differentiated activities from the same worksheet, a list of strategies for differentiated teaching is presented, with some examples of how they can be used in the classroom. The examples can be found at the appropriate places in the text.

1. Using a range of teaching styles
A class is made up of different personalities, who probably have preferred learning styles. Using a range of teaching approaches makes it more likely that all students will be able to respond to the science that is being taught. The following examples have been included and can be found at the appropriate place in the resource.

Example What happens when things burn?
Approach 1 – Working in groups and reporting back to the whole class
Approach 2 – A thinking skills lesson

RS•C

2. Varying the method of presentation or recording
Giving the students some choice about how they do their work. There are many opportunities given throughout the resource.

3. Taking the pupil's ideas into account
Provide opportunities for students to contribute their own ideas to the lesson. For example when setting up an investigation allow different students the freedom to chose which variables they are going to investigate. The use of concept cartoons provides an ideal opportunity for students to discuss different scientific concepts (see D. Warren, *The nature of science*, London: Royal Society of Chemistry, 2001.)

4. Preparing suitable questions in advance
Class discussions are important in motivation, exploring ideas, assessment *etc*. Having a list of questions of different levels prepared in advance can help to push the class.

5. Adjusting the level of scientific skills required
Example – Using symbol equations or word equations

6. Adjusting the level of linguistic skills required
Example Norman Rillieux – Sparkling white crystals of sugar
Sheet 1 – a high reading age
Sheet 2 – a low reading age

Teachers may like to check the readability of their materials and of the texts they use. Guidance on this and on the readability of a range of current texts may be found at **http://www.timetabler.com/contents.html** (accessed June 2001).

7. Adjusting the level of demand on the student
Example
What happens when things burn? – both approaches are for the more able student. Burning theories is much easier to understand and less demanding on students.

References

1. A. Lewis, *British Journal of Special Education*, 1992, **19**, 24–7.

2. S. Naylor, B. Keogh, *School Science Review*, 1995, **77(279)**, 106–110.

How scientists communicate their ideas

Effective communication is crucial to the advancement of science and technology. All around the globe there are groups of research scientists and engineers, in universities and in industry, working on similar scientific and technological projects. Communication between these groups not only gives the scientist new ideas for further investigations, but helps in the evaluation of data. Results from different groups will either help to confirm or reject a set of experimental data. Communication is vital when a company wants to sell a new product. Depending on the product the buyer will want to understand how it works and how to maintain it. Several of the employees will have to learn how to use the product, and respond quickly to changing technology and circumstances. Therefore the manufacturers must be able to communicate the science to prospective buyers.

RS•C

Scientists communicate in a number of ways including:

■ Publication in research journals

■ Presenting papers at scientific conferences

■ Poster presentations at conferences

■ Book reviews by other scientists

■ Publication on the Internet

■ Sales brochures

■ Advertising flyers

■ Television documentaries

Publication in research journals

The article is written. The article must have an abstract, which is a short summary.

It is submitted to a journal.

The article is refereed by other scientists, working in a similar area. This is to check that the work is correct and original.

The article may be returned to the author to make changes.

The article is accepted and published by the journal.

The article is published.

Presenting papers at scientific conferences

Conference organisers invite scientists to speak on specific topics and projects.

An abstract is submitted to and accepted by the conference organisers.

The conference programme is organised and the speakers notified.

The scientist gives their talks, usually aided by slides, which contain the main points.

There is usually time for questions after the talk.

The written paper is given to the conference organisers.

All the papers are published in the conference proceedings. This is usually a book.

Poster presentations at conferences

An abstract is submitted to and accepted by the conference organisers.

The conference programme is organised and the poster people notified.

During the poster session the authors stand by the posters ready to answer any questions, as the delegates read the posters.

Written papers may then be published in the conference proceedings.

Book reviews

Other scientists in the same field often review new books. The reviews are then published in scientific magazines and journals. The review offers a critical summary of the book. The idea of the review is to give possible readers an idea of the contents and whether it is suitable for the intended purpose.

RS•C

Publishing on the Internet

This is the easiest way to publish. Anyone can create their own web page and publish their own work. In this case the work is not refereed or checked by other people.

However, a lot of the information published on the Internet is linked to reputable organisations. In this case the articles will have been checked before they are published. Much of the information published on the Internet is targeted at the general public, and therefore the scientific ideas are presented in a comprehensible way. There are often chat pages so people can communicate their views and ask questions or request further information. The power of the Internet is that there is the opportunity to get immediate feedback to a comment or question.

Sales brochures

The information must be presented in an attractive and concise manner. After all you are trying to sell something. There should be a balance between technical information and operating instructions!

Advertising flyers

This must be written with the target audience in mind.

The information must be concise as there is limited space. The format must be attractive and should include pictures as well as writing. The flyer should also be quite cheap to produce.

Teaching students to communicate ideas in science

Students can be taught effective communication skills:

■ By encouraging communication between students and a range of audiences in classrooms

■ By encouraging them to investigate like 'real scientists' by reporting their findings for checking and testing by others, and participating in two-way communication. (Communicating between groups, classes, partner schools, schools abroad perhaps via the Internet.)

■ By setting investigations in a social context which offers the opportunity to communicate the project outside of the classroom. These work best when there is local interest.

When presenting investigative work to an audience, the student should consider the following:

■ Who will be in the audience?

■ What information does the audience need to know eg method, results and recommendations?

■ How to present the information in an interesting and professional way eg should graphs be hand drawn or done on the computer?

■ That the information offered convinces the audience that their investigation was valid and reliable.

■ Poster presentations or display boards should be concise, since the space is limited.

■ When speaking to audiences remain calm, speak clearly and slowly and try to be enthusiastic. Make sure that information on slides and OHTs can be read from the back of the room.

RS•C

When writing a report of the findings of a scientific investigation for others to check and test, the emphasis should be on clarity. Another person is going to carry out the same investigation. The only information available is what is written in the report.

The report could be written under the following headings:

- Introduction
- Scientific knowledge
- Planning
- Table of results
- Graphs
- Conclusions
- Evaluation
- Recommendations

Further background information

R. Feasy, J. Siraj-Blatchford, *Key Skills: Communication in Science*, Durham: The University of Durham / Tyneside TEC Limited, 1998.

Curriculum coverage

Curriculum links to activities in this resource are detailed at
http://www.chemsoc.org/networks/learnnet/social-hist.htm

Curriculum links to activities in other resources in this series are detailed at

http://www.chemsoc.org/networks/learnnet/ideas-evidence.htm

Health and safety

All the activities in this book can be carried out safely in schools. The hazards have been identified and any risks from them reduced to insignificant levels by the adoption of suitable control measures. However, we also think it is worth explaining the strategies we have adopted to reduce the risks in this way.

Regulations made under the Health and Safety at Work *etc* Act 1974 require a risk assessment to be carried out before hazardous chemicals are used or made, or a hazardous procedure is carried out. Risk assessment is your employers responsibility. The task of assessing risk in particular situations may well be delegated by the employer to the head of science/chemistry, who will be expected to operate within the employer's guidelines. Following guidance from the Health and Safety Executive most education employers have adopted various nationally available texts as the basis for their model risk assessments. These commonly include the following:

Safeguards in the School Laboratory, 11th edition, ASE, 2001

Topics in Safety, 3rd Edition, ASE, 2001

Hazcards, CLEAPSS, 1998 (or 1995)

Laboratory Handbook, CLEAPSS, 1997

Safety in Science Education, DfEE, HMSO, 1996

Hazardous Chemicals – a manual for science education, SSERC, 1997 (paper).

Hazardous Chemicals – an interactive manual for science education, SSERC, 1998 (CD-ROM)

RS•C

If your employer has adopted more than one of these publications, you should follow the guidance given there, subject only to a need to check and consider whether minor modification is needed to deal with the special situation in your class/school. We believe that all the activities in this book are compatible with the model risk assessments listed above. However, teacher must still verify that what is proposed does conform with any code of practice produced by their employer. You also need to consider your local circumstances. Is your fume cupboard reliable? Are your students reliable?

Risk assessment involves answering two questions:

■ How likely is it that something will go wrong?

■ How serious would it be if it did go wrong?

How likely it is that something will go wrong depends on who is doing it and what sort of training and experience they have had. In most of the publications listed above there are suggestions as to whether an activity should be a teacher demonstration only, or could be done by students of various ages. Your employer will probably expect you to follow this guidance.

Teachers tend to think of eye protection as the main control measure to prevent injury. In fact, personal protective equipment, such as goggles or safety spectacles, is meant to protect from the unexpected. If you expect a problem, more stringent controls are needed. A range of control measures may be adopted, the following being the most common. Use:

■ a less hazardous (substitute) chemical;

■ as small a quantity as possible;

■ as low a concentration as possible;

■ a fume cupboard; and

■ safety screens (more than one is usually needed, to protect both teacher and students).

The importance of lower concentrations is not always appreciated, but the following table, showing the hazard classification of a range of common solutions, should make the point.

Ammonia (aqueous)	irritant if ≥ 3 mol dm^{-3}	corrosive if ≥ 6 mol dm^{-3}
Sodium hydroxide	irritant if ≥ 0.05 mol dm^{-3}	corrosive if ≥ 0.5 mol dm^{-3}
Ethanoic (acetic) acid	irritant if ≥ 1.5 mol dm^{-3}	corrosive if ≥ 4 mol dm^{-3}

Throughout this resource, we make frequent reference to the need to wear eye protection. Undoubtedly, chemical splash goggles, to the European Standard EN 166 3 give the best protection but students are often reluctant to wear goggles. Safety spectacles give less protection, but may be adequate if nothing which is classed as corrosive or toxic is in use. Reference to the above table will show, therefore, that if sodium hydroxide is in use, it should be more dilute than 0.5 M (M = mol dm^{-3}).

CLEAPSS Student Safety Sheets

In several of the student activities CLEAPSS student safety sheets are referred to and recommended for use in the activities. In other activities extracts from the CLEAPSS sheets have been reproduced with kind permission of Dr Peter Borrows, Director of the CLEAPSS School Science Service at Brunel University.

RS•C

- Teachers should note the following points about the CLEAPSS Student Safety Sheets:

- Only extracts from fuller student safety sheets have been reproduced.

- Only a few examples from a much longer series of sheets have been reproduced.

- The full series is only available to member or associate members of the CLEAPSS School Science Service.

- At the time of writing, every LEA in England, Wales and Northern Ireland (except Middlesbrough) is a member, hence all their schools are members, as are the vast majority of independent schools, incorporated colleges and teacher training establishments and overseas establishments.

- Members should already have copies of the sheets in their schools.

- Members who cannot find their sheets and non-members interested in joining should contact the CLEAPSS School Science Service at Brunel University, Uxbridge, UB8 3PH; tel. 01895 251496; fax. 01895 814372; email science@cleapss.org.uk or visit the website **http://www.cleapss.org.uk** (accessed June 2001).

- In Scotland all education authorities, many independent schools, colleges and universities are members of the Scottish Schools Equipment Resource Centre (SSERC). Contact SSERC at St Mary's Building, 23 Holyrood Road, Edinburgh, EH8 8AE; tel. 0131 558 8180, fax 0131 558 8191, email sts@sserc.org.uk or visit the website **http://www.sserc.org.uk** (accessed June 2001).

RS•C

RS•C

Using the resource on an Intranet

If your school or college has an Intranet you may wish to download the material in part or in whole to the Intranet to facilitate easy links for students. Alternatively you could use some of the web references given to design an interactive worksheet. Instructions for this are given below.

Designing an interactive worksheet
These instructions are for Microsoft Word.

■ First you need to do some research on the Internet to find the relevant sites related to the topic and note down the website addresses, or use the sites already known and available from the resource.

■ Choose an appropriate font, size and colour for your text.

■ Type in the title of the page.

■ Save this page as an HTML file in order to make the page 'live'.

■ Type in the instructions and questions that you want the students to read.

■ Type in the web page address (url). Always begin http://www (this will automatically make the page a hyperlink to the website you have typed and the text will turn blue).

Now you can make your work look more like a web page by placing lines, graphics, scrolling text and backgrounds in it.

■ To place a horizontal line on your page, put the cursor where you would like the line to be, then click on insert, go to horizontal line, choose a line and click on OK.

■ To add a background to your page, click on format, go to background, then fill effects option, choose an effect and click on OK.

■ To add a picture to your work, place the cursor where you would like the picture to be, then click on insert, go to picture and clip art option. Choose a suitable picture for your work and click on OK. The picture size can be altered by moving the edges in or out or the picture can be moved to another place by dragging it over.

■ To place scrolling text in your work, you need to highlight the words, click on insert, go to scrolling text, choose background colour, the speed of the scroll and press OK. (If you are going to print this work out, the scrolling text will not print out).

Your work is now beginning to look like an interactive web page.

■ When you are satisfied with the final product, click on file, go to web page preview and this will show you what your page looks like. (You cannot alter your page through the web page preview screen; you will have to go back to Word).

■ Remember to do a spell check on your work, *ie* click on the ABC icon on the top of toolbar.

■ Test out your page.

Now you are ready to use the page with students.

RS•C

This page has been intentionally left blank.

RS•C

Introduction

Chemistry has a human face. The aim of this book is to present chemists as real people and not stereotypical 'mad scientists' whose lives are completely dominated by science. It may only take a couple of minutes of a lesson to present the class with a bit of personal background information which could, for some students, add interest to the lesson. The history of chemistry is full of serendipitous tales. The influence of World War II played a major role in the development of plastics, which otherwise may not have been produced on a commercial scale. In other parts of the world science was held back, simply because the scientists were coloured and this was socially unacceptable. In the early days of chemistry, there were few woman chemists as this too was deemed to be unacceptable. The womans' place was in the home and certainly not in the laboratory carrying out experiments. The social context of the time must be understood to explain why the majority of the early chemists appear to be white middle-class or upper-class gentlemen. Many of the scientific projects had to be funded by the chemists themselves, so they already had to be wealthy or have a very good income. At times science was just the hobby, before it took over. This is not to say that there were no scholarships available. In fact, there are a number of very distinguished scientists such as Ernest Rutherford and John Jacob Berzelius who came from very humble backgrounds.

Today we live in a very different world, where people should be accepted regardless of race or gender. Science is taught in all schools and young people of today are encouraged to develop an interest in the subject. There is more money available to carry out research, although some would say not enough, as we go on living in the scientific and technological age.

Running throughout this series of books written by Dorothy Warren, there are many references to different scientists. The scientists are always introduced within the context of their work at the most appropriate place; for example John Lind is found in **The nature of science** book, Alice Hamilton in the **Health, safety and risk** book, Mario Molina in the **Climate change** book and the aluminium pioneers in the **Green Chemistry** book. In this resource, the focus is twofold, namely providing strategies for teaching about people in chemistry and an introduction to some of the chemists who played a role in the development of major ideas in chemistry, *eg* theories about the atom and burning. Roy Plunkett and the discovery of Teflon has been included as an example of serendipity, Harry Kroto and buckminsterfullerene as an example of a living chemist and Norbert Rillieux as an example of a successful chemical engineer, despite being an African-American living in the 19th century.

One of the problems for busy teachers is having a readily available source of background information about different scientists. There are many Internet sites that will provide a wealth of biographical information, photographs and scientific information. The three web sites listed here are worth a visit.

1. The British Society for the History of Science (BSHS) website
   http://www.man.ac.uk/Science_Engineering/CHSTM/bshs/ (accessed June 2001).
   This site links into several other useful sites. It also provides a discussion forum for teachers as well as useful contacts such as actors willing to do scientific performances in schools.

RS•C

2. The Nobel Foundation site at **http://www.nobel.se** (accessed June 2001) lists all the Nobel Laureates with photographs, biographical and scientific background information.

3. This week in the history of chemistry **http://webserver.lemoyne.edu/faculty/giunta/week.html** (accessed June 2001) can be accessed to provide information for any week of the year.

4. European Network for Chemistry, Millennium Project, has a site listing 100 distinguished European chemists from the chemical revolution to the 21st century. **http://www.chemsoc.org/networks/enc/fecs/100chemists.htm** (accessed June 2001).

5. Chemsoc timeline is a linear based exploration of key events in the history of science with a particular emphasis on chemistry. **http://www.chemsoc.org/timeline** (accessed June 2001).

RS•C

RS•C

The atom detectives

Teachers' notes

Objectives
- To understand how the model of the atom has developed over time.
- To learn about some of the chemists involved in developing the model of the atom.
- To be able to apply today's accepted model of the atom and draw diagrams to represent the atoms of the first 40 elements of the periodic table.

Outline
This section looks at how the model of the atom has developed over the last 200 years. It explores how scientists work together to develop new ideas and how new theories may, at first, give rise to controversy. It shows how technological advances can lead to the development of new theories and ideas.

Teaching topics

This material is intended to be used with more able students between the ages of 14 and 16 or post-16 students, when teaching about atomic structure and the Periodic Table. Understanding the arrangements of subatomic particles in the atom is a high level concept. In this activity, you can see how the ideas developed as scientific method and instrumentation developed and other areas of science unfolded. Look at the student worksheets before reading the detailed notes below.

Background information

Since 5 BC, people have been curious to find out more about different materials and substances. The theory of Democritus said 'Substances are different because homogenous particles have different sizes and shapes and cannot be cut'. This was just the start of many more theories that would be put forward and then rejected over the next 2000 years. Many of the early theories of matter were not based upon experiments. As scientists began to study the relationship between physical phenomena such as electricity and magnetism they began to develop different models about atomic structure.

Sources of information

The atomic structure timeline at
http://www.watertown.k12.wi.us/hs/teachers/buescher/atomtime.asp
(accessed June 2001).

The timeline has twenty-two entries, starting in the Greek era and finishing in 1932 with James Chadwick. By clicking on the scientist's name, personal background information and portraits can be accessed. A short summary of their contribution to atomic structure is included with some hot links leading to further information.

Teaching tips

This section is ideal for group work leading to a wall display featuring the history of the atom. There are five student worksheets each featuring a scientist who made a significant contribution to the development atomic theory.
- Dalton
- Berzelius
- Thomson

RS•C

- Rutherford
- Bohr.

The sixth student worksheet is an information sheet, bringing the theory up-to-date.

Following an introduction to the lesson, the class could be divided into groups of 4 or 5 students. Each group could work through one of the worksheets. To make the sheets more durable, they could be photocopied onto card and laminated. The activities at the bottom of the sheet could be carried out and presented as a poster, on a large piece of paper. The group should also include some background information about the scientists and what they did.

You will need to tell the class whether you expect models or diagrams of the atoms, or if they have a free choice. In order to answer all the questions some groups will have to communicate with others.

At the start of the next lesson, each group could present their posters, in chronological order, to the rest of the class, highlighting the major aspects of the theory, which could be summarised.
Eg 1 Berzelius' relative atomic masses are used today. Berzelius introduced chemical symbols.
Eg 2 Thomson discovered that the negatively charged electron was part of the atom. It was 1/1840 the mass of the positively charged particles.

The rest of the lesson could be spent with everyone working on the sheet **Modelling the atom today**. You should go through the lithium example with the class and maybe do one of the other examples with the class. The information for completing question 1 will have been covered in the poster presentations and may already be available. The rest of the sheet could be completed as homework.

Note: Throughout this material, the term mass has been used unless the accepted terminology of weigh or weight is more appropriate. Teachers may wish to draw attention to the difference between mass and weight.

Resources

- Student worksheets
  – John Dalton
  – John Jacob Berzelius
  – Joseph John Thomson
  – Ernest Rutherford
  – Niels Bohr
  – Present day models of the atom
  – Modelling the atom today

Timing

2 hours plus homework

Opportunities for using ICT

- Using the Internet to find out more about the atom detectives.
- Word processing and drawing packages to make posters.

Opportunities for using key skills

- Working together in groups.
- Communication between groups.
- Presentation of work to the class.

RS•C

Answers

John Dalton

1.

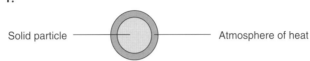

Solid particle ——————— Atmosphere of heat

2. A molecule of ice had less heat surrounding the particle than a molecule of water.

3. He guessed!

4.

Water · · · · · · · · · · · · · or · · · · · · · · · · · · · · · · · · · CO_2

Dalton thought that the water molecule only consisted of one O and one H atom, so either molecule may be accepted.

John Jacob Berzelius

1.

Element	Symbol	Relative atomic mass
Chlorine	Cl	35.5
Copper	Cu	63.5
Hydrogen	H	1.0
Lead	Pb	207.2
Nitrogen	N	14.0
Oxygen	O	16.0
Potassium	K	39.1
Silver	Ag	107.9
Sulfur	S	32.1

2. Berzelius because his 1826 values are very close to the ones that are used today.

3. He was able to work out the number of atoms in each molecule.

4. Berzelius believed that atoms were held together by electrostatic attraction. He thought that some atoms were positively charged and others were negatively charged.

Joseph John Thomson

1.

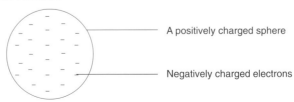

————— A positively charged sphere

————— Negatively charged electrons

The plum pudding model

2. Thomson's model of the atom contained negative particles of electricity (which he called electrons) embedded in a solid sphere of positive charge. Daltons' atom was a solid particle, surrounded by an atmosphere of heat.

3. A very sensitive camera.

Ernest Rutherford

1.

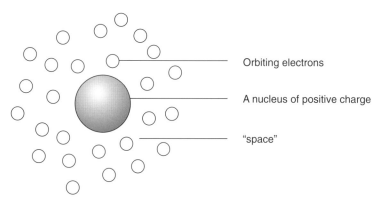

Orbiting electrons

A nucleus of positive charge

"space"

2. The main difference between the two models is that the Thomson model has a solid sphere with negative charges whereas the Rutherford model has a small solid nucleus, some 'space' and then orbiting electrons, which are separate particles.

3. Rutherford carried out his experiments in the dark so that he could observe the glow left behind by the radiation and see where the particles went.

4. Rutherford noticed that some of the positively charged radiation bounced back from the atom in the same direction. Rutherford concluded that this must be due to another positive force repelling the positive radiation. This large positive force must come from the centre of the atom.

Niels Bohr

1.

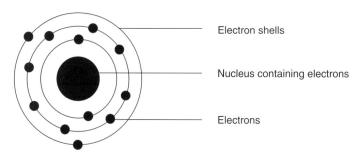

Electron shells

Nucleus containing electrons

Electrons

2. In the Bohr atom the electrons are arranged in definite shells whereas in the Rutherford model, the electrons just orbit the nucleus. They are free to go wherever they like.

3. Bohr's model of the atom was based on theoretical calculations and a good imagination. Although his model could explain atomic spectra, it was based on incomplete data. It was thought that the model could not be used to explain the reactivity of the elements. Many chemists preferred Lewis' model because it was based on real experimental chemical data. The Lewis model could be used to explain the reactivity of the elements. However, it could not be used to explain the hydrogen spectrum.

Modelling the atom today

1.

Particles	Charge	Relative Mass
Protons	positive (+)	1
Neutrons	neutral	1
Electrons	negative (–)	1/1840

2.

Element	Mass No.	Atomic No.	No. of protons	No. of neutrons	No. of electrons
Hydrogen	1	1	1	0	1
Carbon	12	6	6	6	6
Neon	20	10	10	10	10
Aluminium	27	13	13	14	13
Potassium	39	19	19	20	19

3. Nucleus containing correct number of protons and neutrons (see table), electrons arranged in shells 2,8,8,18 *etc.*

John Dalton

John Dalton
(Reproduced courtesy of the Library &
Information Centre, Royal Society of
Chemistry.)

John Dalton (1766–1844)

He was the son of an English weaver from Eaglesfield in Cumbria. When he wasn't carrying out investigations, he was probably teaching at the Presbyterian college in Manchester. In 1807, John Dalton was the first person to use the word **atom** to describe the smallest particle of any element.

What did Dalton do?

Dalton studied gases and discovered that elements combine with other elements to make compounds. He had to guess how many atoms joined together to make the compound. He was able to calculate the relative weights of particles using data from his own observations and measurements. Individual particles were too small to weigh. He had to make some assumptions to explain his observations *eg* the atmosphere of heat surrounding the solid particle was used to explain why some elements were solids and some gases. Solid compounds had less heat than gaseous ones.

Dalton's atomic theory of matter, 1807

1. All matter is made up of tiny particles called atoms.

2. Each atom is a solid particle with no spaces, surrounded by an atmosphere of heat.

3. Atoms cannot be made or destroyed.

4. Atoms of the same elements are alike with the same mass, colour *etc.*

5. Atoms of different elements have different masses, colours *etc.*

6. Atoms can join to form larger particles in compounds.

○ Oxygen

 Hydrogen

 Carbon

Dalton's symbols

Things to do

1. Make a model or draw a diagram of Dalton's atom.

2. How did Dalton explain the different between a molecule of ice and a molecule of water?

3. How did Dalton known how many atoms were in a molecule?

4. Using Dalton's symbols, write down the formulae of water and carbon dioxide.

RS•C

John Jacob Berzelius

John Jacob Berzelius
(Reproduced courtesy of the
Nobel Foundation.)

John Jacob Berzelius (1779–1848)

Berzelius was an orphan from Sweden, brought up by a mean stepfather. He worked on his farm and lived in a room which was also the potato store. His stepfather made sure that the potatoes did not freeze during winter, so at least this meant that Berzelius kept warm too. From high school, he went on to university where he became interested in experimental chemistry.

What did Berzelius do?

Berzelius heard about Dalton's theory and set about making his own relative atomic weight measurements. But, from previous experiments carried out by Humphrey Davy, he knew how many atoms were in the compounds. He knew that when electricity was passed through water, twice as much hydrogen was collected at the negative terminal than oxygen at the positive terminal. So he concluded that water was made from two atoms of hydrogen and one of oxygen.

Berzelius' atomic theory

1. All atoms are spherical.

2. All atoms are the same size.

3. Atoms have different weights.

4. Atoms joined together in fixed proportions, by an electrochemical reaction. Some atoms are positive and others are negative.

Dalton could not accept Berzelius' electrochemical combination, but at the same time could not explain why atoms joined together in fixed proportions.

Elements	Dalton's atomic weights 1808	Berzelius' atomic weights 1826
Chlorine	unknown	35.41
Copper	56	63.00
Hydrogen	1	1.00
Lead	95	207.12
Nitrogen	5	14.05
Oxygen	7	16.00
Potassium	unknown	39.19
Silver	100	108.12
Sulfur	13	32.18

Chemical symbols

Berzelius thought that chemical symbols should be letters. He took the first letter of the Latin name of each element. When the letters were the same, he used both the first letter and the next different letter. Berzelius' symbols are used in today's Periodic Table.

Things to do

1. Look up, in a modern data book, the relative atomic masses and the symbols of the elements listed in the above table. Put your answers in a table.

2. Who do you think had the best method for calculating the relative weights of atoms, Dalton or Berzelius?

3. What do you think was the key to the successful calculations?

4. How did Berzelius think that atoms joined together to make compounds?

Joseph John (JJ) Thomson

Joseph John Thomson
(Reproduced courtesy of the
Nobel Foundation.)

Joseph John Thomson 1856–1940

Thomson was born near Manchester. His ambition was to be an engineer but instead he was awarded a scholarship in chemistry. The scholarship was in memory of John Dalton. At the age of 28, Thomson became professor at the Cavendish Laboratory, Cambridge University.

What did Thomson do?

In 1897 Thomson discovered the electron, while he was investigating the conductivity of electricity by gases at very low pressures. After collecting data for twenty years, Thomson was convinced that electrons were negative particles of electricity. He even measured the mass of the electron.

However, he still needed more evidence to convince the scientific world, so he asked Wilson to try and take a photograph of an electron. It took him until 1911 to build a suitable camera, which was sealed in a glass chamber in which electrons could be produced. The experiment was successfully carried out and the electron was photographed.

JJ Thomson was worried about telling the world his new theory of the atom, because until now the atom was thought of as a single solid particle.

Thomson's model of atomic structure – 1899

- Atoms consisted of rings of negative electrons embedded in a sphere of positive charge (the plum pudding model).

- The positive and negative charges balance to make the atom neutral.

- The mass of the atom was due to the nucleus.

- The mass of an electron was 1/1840 of the mass of hydrogen, the lightest atom.

- There were 1840 electrons in an atom of hydrogen.

Things to do

1. Make a model or draw a diagram of JJ Thomson's model of the atom.

2. What is the main difference between this new model and Dalton's model?

3. What advances in technology made it possible for Thomson to successfully complete his investigations?

Ernest Rutherford

Ernest Rutherford
(Reproduced courtesy of the
Nobel Foundation.)

Ernest Rutherford (1871–1937)

Rutherford was born near the village of Nelson, New Zealand. His father was an odd-job man and simple farmer. Rutherford obtained an honours degree in mathematics and science from the University of New Zealand before gaining a scholarship that took him to work with JJ Thomson at the Cavandish laboratory in Cambridge.

What did Rutherford do?

Rutherford studied radioactive atoms and found that they were not stable. By this time a lot was understood about radiation. Rutherford carried out his investigation in the dark. He used positively charged radiation to bombard the atom and watched to see where the radiation particle went. The radiation always left a glow. The glow showed that most particles went straight through the atom, some were slightly deflected, while others bounced back in the same direction.

After doing many calculations, Rutherford concluded that the radiation could only come back if that atom had a hard positively charged core at the centre of the atom. He called this the nucleus. If the atom was 100 m, the size of a football pitch, the nucleus would be the size of a pea placed in the centre of the pitch.

Rutherford's model of atomic theory

1. The atom consists mainly of space.

2. The mass of the atom is concentrated in the nucleus, which is a small core at the centre of the atom.

3. The nucleus has positive charges.

4. Electrons move around the nucleus like planets orbiting the sun.

5. The atom is neutral as it has the same number of positive charges and negatively charged electrons.

Things to do

1. Make a model or draw a diagram of Rutherford's atom.

2. What was the main difference between Rutherford's model of the atom and Thomson's model?

3. Why did Rutherford carry out his experiments in the dark?

4. What evidence do you think lead Rutherford to conclude that the atom had a positively charged nucleus?

Niels Bohr

Niels Bohr
(Reproduced courtesy of the
Nobel Foundation.)

Niels Bohr (1885–1962)

Niels Bohr was born into a scientific family. His father was a professor of physiology and his brother a distinguished mathematician. After obtaining his Ph.D. from the University of Copenhagen, Denmark, he accepted an invitation to work with Rutherford, at Cambridge.

Bohr was very intelligent and had an amazing imagination. He was not afraid to build on the idea of Max Plank, that energy came in little packets called quanta, and apply this to Rutherford's model of the atom.

What helped Bohr?

Bohr based his investigation on Max Plank's idea. He imagined the electron orbiting the nucleus unless it was disturbed by some outside force, when it jumped to a different energy level. A packet of energy was either gained or lost.

Bohr's model of the atom (1922)

1. Most of the mass of an atom is in the central nucleus that contains protons.

2. The electrons are arranged in definite shells or energy levels and orbit the nucleus.

3. The electron shells are a long way from the nucleus.

4. When one shell is full a new shell is started. This is called the electronic configuration.

5. Atoms with full shells are not very reactive.

6. Electrons determine the reactivity of the atom.

Chemists not happy

While Bohr's model of the atom could explain the spectrum of the hydrogen atom, chemists didn't think it would explain the reactivity of the other chemical elements. His theory was based on incomplete physical data and mathematical calculations. Many chemists favoured Lewis' theory. The American's octet theory was based on real chemical data. Lewis proposed that the fixed nucleus was surrounded by cubic shaped electron shells. The electrons were fixed in the corner positions.

Things to do

1. Make a model or draw a diagram of Bohr's atom.

2. What is the main difference between Bohr's model of the atom and Rutherford's model?

3. Suggest why some chemists preferred Lewis' model of the atom to Bohr's.

RS•C

Present day models of the atom

The up-to-date model of the atom is much more complicated than the ones we have met so far. It is based on what is known as quantum mechanics and the atom is described by a complicated equation called the Schrodinger equation. It helps to know a lot of maths, understand probability and have a good imagination to picture Schrodinger's model of the atom. However, you don't need to worry about that until you go on to further studies of chemistry or physics. Of course, if you are interested to find out more you can go and look it up yourself.

Today, most people are happy to accept a slightly modified model of Bohr's atom, because it can be used to explain the spectra and the reactivity of the elements. The modified model of the Bohr atom includes neutrons, which were discovered by Sir James Chadwick in 1932.

Figure 6 James Chadwick
(Reproduced courtesy of the Nobel Foundation.)

The neutron is a neutral particle with the same mass as a proton. It is also found in the nucleus.

Thanks to the work of many other chemists, we can now use the Periodic Table to find out the number of protons, neutrons, and electrons in the atom of any element.

Modelling the atom today

Study the following example

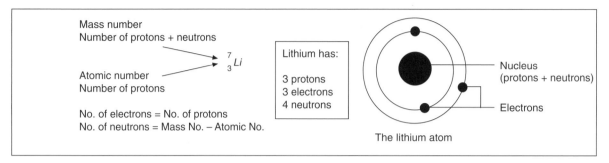

Where the electrons go

■ The first shell can hold up to 2 electrons.

■ The second and third shells can each hold up to 8 electrons.

■ The fourth shell can hold up to 18 electrons.

The arrangement of electrons in the atom is known as the electronic configuration. The electronic configuration for Li is 2.1. This means Li has 2 electrons in the first shell and one in the second shell.

Things to do

1. Complete the following table

Particles	Charge	Relative Mass
Protons		1
Neutrons	neutral	
Electrons		

2. Complete the following table.

Element	Mass No.	Atomic No.	No. of protons	No. of neutrons	No. of electrons
Hydrogen	1	1			
Carbon			6	6	
Neon	20		10		10
Aluminium		13		14	
Potassium	39	19			

3. For each of the elements in the table above, draw a diagram to show how the protons, neutrons and electrons are arranged in each atom.

RS•C

RS•C

What happens when things burn?

Teachers' notes

Objectives
■ To understand that today's accepted theory of burning is the combination of oxygen with other substances.

■ To know that in the past many scientists believed the Phlogiston theory.

Outline
The investigation **What happens when things burn?** is intended to be a group activity that will investigate the 'Phlogiston Theory' of burning put forward in 1732 by George Stahl. Through making predictions, carrying out experiments, discussions and researching information, the students must decide whether to accept or reject the theory. Their decision must be supported by evidence. If the theory is rejected, a new theory of burning must be put forward.

Two different approaches to carrying out this investigation have been included.

■ **Approach 1.** Different groups carry out different experiments and then pool the results.

■ **Approach 2.** A thinking skills lesson, in which every student sees all the experiments.

Burning theories investigates burning, before looking back to see what 18th century chemists believed.

Teaching topics

The activity **What happens when things burn?** is suitable for high ability 13–14 year olds. It is not suitable for less able students who may find the investigation too confusing.

Burning theories is suitable for most 13–14 year olds.

Either activity could be included when teaching about burning and the formation of compounds.

Look at the student worksheets before reading the rest of the teachers' notes.

Sources of information

M. E. Bowden, *Chemical Achievers: the human face of the chemical sciences*, Philadelphia: Chemical Heritage Foundation, 1997.

H. W. Salzberg, *From Caveman to Chemist: Circumstances and Achievements*, Washington: American Chemical Society, 1991.

W. H. Brock, *The Fontana history of chemistry*, London: Fontana Press, 1992.

B. Jaffe, *Crucibles: The story of Chemistry* (Fourth Revised Edition), New York: Dover Publications,1976.

RS•C

Approach 1: What happens when things burn?

Teaching tips

Lesson 1

Introduce the class to Phlogiston theory by handing out the student information sheets, **What happens when things burn?** and **About the chemists** and explain the task. This activity is suitable for high ability 13–14 year old students.

Task

The task is to test out the Phlogiston theory and decide whether to accept it or reject it. The enquiry involves the students collecting data from a variety of sources such as doing their own experiments, finding out the results from other groups, and using **Burning – the fact cards**.

At the end of the enquiry, the students must come up with a firm conclusion, which should include their reasoning.

First demonstrate experiment B, smelting. Students could fill out the sheet while you do this.

Divide the class in half. One half could carry out experiment A in pairs, while the other half does experiment C.

Lesson 2

Hand out the **Burning interpretations table**. A few groups could present their results so that everyone may complete the interpretations table. The results should be discussed and they should decide if there is enough evidence to either support or reject the theory. State explicitly that a scientific theory is only valid if all the evidence fits it – if not we need to look for another theory.

It should be suggested that further evidence is still needed as the results are not conclusive. A possible way forward would be to investigate the work that other scientists, such as Black, Priestley and Lavoisier did (all are mentioned on the student information sheet). They should be looking for experiments in which samples were weighed before and after heating.

Hand out **Burning – the fact cards** and discuss the information in groups.

Some teachers may wish the class to find out their own information about burning theories and the people who thought them up. Many science CD-ROMS can be used for this.

By the end of the lesson each group must come to a firm conclusion, either accepting or rejecting the Phlogiston theory. They should be able to write a report stating their conclusion, explaining how they came to it, and where applicable putting forward a new theory.

In the final section of the lesson, today's accepted theory of combustion should be covered. It is important to make sure that the students do not go away with any new misconceptions. If there is time, go through the chemistry of experiments A-C. It is also worth pointing out that even though Priestley showed by experiment that metals combine with oxygen when they burn, the exact opposite of what the Phlogiston theory said, he still believed the Phlogiston theory when he died. He could not bring himself to reject such a well known theory even though it was completely wrong. Also note that instrumentation was not very good in the 18th century and not all chemists could agree if mass was lost or gained during the reactions.

Resources

- ■ Class sets of student worksheets
  - What happens when things burn? (1)
  - About the chemists

RS•C

- Experiment A – Calcination
- Experiment B – Smelting
- Experiment C – Making alkali
- Burning interpretations table
- Burning – the fact cards

Experiment A (per group)
- Balance (measuring to 2 decimal places if possible)
- Magnesium ribbon 10–20 cm
- Heating crucible and lid
- Bunsen burner
- Heatproof mat
- Tripod
- Pipe clay triangle
- Tongs
- Safety glasses

Experiment B (Demonstration)
- Balance (measuring to 2 decimal places if possible)
- Combustion tube
- Bung fitted with a small glass tube
- Porcelain boat
- Copper(II) oxide
- Methane (natural) gas
- Bunsen burner
- Safety screen

Experiment C (per group)
- Balance (measuring to 2 decimal places if possible)
- Limestone chips
- Wire ring, or gauze (place the limestone chip near the edge and heat)
- Bunsen burner
- Heatproof mat
- Tongs
- Safety glasses
- Nails
- Watch glasses
- Dropper
- Universal Indicator paper or litmus paper.

Practical tips

Experiment A
Demonstrate the technique of lifting the lid a little, to allow access for the oxygen. The purpose of the lid is to minimize any loss of powder as smoke.

Students must wear safety glasses.

Experiment B
A pin hole test tube can be used and the experiment can be carried out without a porcelain boat.

RS•C

Experiment C

A sample of soft chalk (calcium carbonate) reacts better than a marble chip or limestone. Blackboard chalk is not always calcium carbonate. For this reaction to work well, the heat must be concentrated on the calcium carbonate. Balancing the chip on a small deflagrating spoon gives very good results.

Timing

2 hours 30 minutes for the whole activity.

Approach 2: What happens when things burn?

Teaching tips

In this approach the emphasis is on group work and discussions to enable the students to think through the difficult concepts. It is important that each student sees all the experiments so that they can process the information.

Hand out the student information sheets **What happens when things burn? (2)** and **Testing the Phlogiston theory**. Students should read through the sheets, complete the predictions table and then set up experiment C Making alkali.

While the limestone is being heated, demonstrate experiments A and B. It is useful to discuss the experiments while they are being demonstrated and write the start and finish masses on the board.

Students should return to experiment C and complete it.

The results table and questions 1–5 should be completed.

Students should discuss their results and try and complete the **Burning interpretations table**.

Hand out **Burning – the fact cards** to help students come up with an alternative model of burning.

In the final section of the lesson today's accepted theory of combustion should be covered. It is important to make sure that the students do not go away with any new misconceptions. If there is time, go through the chemistry of experiments A–C. It is also worth pointing out that even though Priestley showed by experiment that metals combine with oxygen when they burn, the exact opposite of what the Phlogiston theory said, he still believed the Phlogiston theory when he died. He could not bring himself to reject such a well-known theory even though it was completely wrong. Also note that instrumentation was not very good in the 18th century and not all chemists could agree if mass was lost or gained during the reactions.

Resources

- As approach 1 for experimental apparatus.
- Student worksheets
  – What happens when things burn? (2)
  – Testing the Phlogiston theory
  – Burning interpretations table
  – Burning – the fact cards

Practical tips

As approach 1

Timing

2 hours

RS•C

Burning theories

Teaching tips

Demonstrate to the class that fuels need oxygen to burn and this oxygen is taken from the air.

Students should then carry out their own experiment to find out what is made when a candle is burnt. More able students should be encouraged to use particle theory to help explain combustion.

Once you are sure that the students understand burning, explain that in the past scientists had a different view of burning. Introduce Joseph Priestly and Antoine Lavoisier as two chemists who played a major role in investigating burning. Their work led to the rejection of the Phlogiston theory and the acceptance of today's theory.

Resources

Demonstration
- Washing up bowl
- Gas jar
- Candle (15 cm)
- Plasticene

Class experiment (per group)
- Gas jar and lid
- Candle on a tray
- Heat-proof mat
- Limewater (0.02 mol dm^{-3})
- Blue cobalt chloride paper
- Safety glasses
- Student worksheets
  – Burning theories

Practical tips

Make sure that the candle is long enough to avoid the top getting wet. This will allow you to repeat the experiment if necessary.

Timing

1 hour

Adapting resources

Approach 1 and Approach 2 of **What happens when things burn?** are an example of how material can be adapted to suit the preferred learning style of the class.

Burning theories is an example of how similar material can be covered in a much less demanding way.

RS•C

In all three cases the learning objectives are the same, and the materials could be modified further to meet the needs of individuals.

Opportunities for using ICT

■ Data logging sensors to measure quantities such as humidity, temperature, light and oxygen levels can be used to monitor the burning process in the Burning theories demonstration.

■ Use of CD-ROMs and the Internet to research 18th century scientists.

■ Word processing to write up the enquiry.

Opportunities for key skills

These activities access the key skill of working together.

Communication skills are important in Approach 1, because each student will not complete all the experiments, but they will need to use the results.

Answers

Approach 1: What happens when things burn?

Experiment A – calcination

1. (a) The mass will decrease as Phlogiston is given off.
   (b) It will become a powder as Phlogiston is given off.

2., 3. –

4. 1(a) – no, the mass has increased, 1(b) yes, a white powder has been formed.

5. 1(a) rejects the Phlogiston theory while 1(b) supports the theory.

6. Yes. Try repeating the experiment using a different metal such as copper, or repeat using a different amount of the metal to see if it still increases.

7. Keep heating and reweighing until there is no further change in mass.

Experiment B – smelting

1. It will turn into a metal because Phlogiston will be transferred from the natural gas to the ore. The metal ore will increase in mass because Phlogiston has been added to it.

2., 3. –

4. Tiny specks of copper have been made.

5. The results support the Phlogiston theory as a metal is made, but they reject the theory because the mass has decreased.

6. Yes, this is only one ore, it needs to be confirmed by others.

7. Phlogiston is a substance that can 'neutralise' the oxygen in the copper ore. It could be carbon.

Experiment C – making alkali

1. It will change into quicklime (CaO) which is alkaline because it will pick up Phlogiston from the air. This will mean that it loses some mass.

2., 3. –

4. Yes, the mass, appearance and pH have all changed.

5. It supports the Phlogiston theory because it has turned into alkali, but it rejects the theory because its mass has decreased.

6. Yes.

7. Try heating another carbonate such as magnesium or copper carbonate.

Completed burning interpretations table

	Prediction using phlogiston theory	Experimental results	Interpretation supports or rejects phlogiston theory
A	Will form a powder Mass decreases	Powder formed Mass increases	Supports theory Rejects theory
B	A metal will form Mass increase	Metal formed Mass decreased	Supports theory Rejects theory
C	An alkali will form Mass increase	Alkali, pH = 11 Mass decreases	Supports theory Rejects theory

An interpretation of Burning - the fact cards

Card	Prediction using Phlogiston theory	Experimental results	Support for or rejection of Phlogiston theory
Black 1	It gains mass as it absorbs Phlogiston	Magnesia loses mass when heated	Rejects
Black 1	The product is alkali	The product is insoluble in water, not alkaline	Rejects
Black 2	It gains mass as it absorbs Phlogiston	Limestone loses 'fixed air'	Rejects
Black 2	It gains mass as it absorbs Phlogiston	Magnesia lost 'fixed' air	Rejects
Priestley 1	No reaction	A gas and liquid is produced	Rejects
Priestley 2		Discovers 'dePhlogistonated' air	
Lavoisier 1	Metals lose mass when they are heated	Metals gain mass when heated	Rejects
Lavoisier 2		Metals mass increases by the amount of oxygen they take from the air	Rejects
Lavoisier 3&4			New theory that explains the mass problems

RS•C

What happens when things burn? (2)

1. Fire is released, water and air escape, the earth or ashes are left behind.

2. Over 2000 years.

3. Particles of fire got stuck between the particles of the material being burnt.

4. Every material that burns has two parts, Phlogiston and ash. When materials burn, they give off Phlogiston.

5. It would make candles burn brightly and would keep mice alive twice as long as in ordinary air.

6. When things burn they join up with oxygen.

7. He knew that when metals burnt their mass increased. The Phlogiston theory said the opposite.

Approach 2: Testing the Phlogiston theory

1. There are several faults in the explanations. In smelting, you would expect both the carbon and the metal ore to give off Phlogiston as the theory says when things burn they give off Phlogiston. In making alkali, why should limestone pick up Phlogiston from the fire. Shouldn't it be losing it?

2. Decrease

3. Weigh the sample, heat it and weigh it again. Continue until there is no change in the mass.

4. Repeat the experiment and take an average of the results.

5. Yes, the mass, appearance and pH have all changed.

Burning interpretation table
See the interpretations table in Approach 1.

Burning theories

1. Level higher than at start.

2. Oxygen

3. Water, carbon dioxide, light and heat.

4. Fuel + oxygen → carbon dioxide + water

RS•C

What happens when things burn? (1)

The ancient greeks thought everything was made from fire, water, air and earth.

So what happens when something burns?

The fire is released.
The water and air escape.
The earth or ashes are left behind.

In the 1600s, scientists thought that burning depended on air and that air was one single substance.

Robert Boyle heated some tin in a sealed flask. He found the tin weighed more after heating than before.

He thought the particles of fire had lodged between the particles of tin.

Georg Stahl (1660 – 1734) was a German scientist who developed another idea called the Phlogiston theory (from the Greek phlox = flame).

Every substance that burns has 2 parts – ash and PHLOGISTON.

When something burns the PHLOGISTON escapes and the ash is left behind.

That's because it contains so much PHLOGISTON!

But this burnt charcoal has only left a little ash.

So it should get lighter.
And you see a powder or ash. This ash is alkaline.

Deciding what happens when things burn

Imagine that you are a scientist in the 1780's and you are fascinated by burning. You want to understand what really happens when something burns. Your favourite explanation is the theory put forward by Georg Stahl in 1723.

When materials burn they give off a substance called Phlogiston.

According to the theory 'Phlogiston' is a real substance with mass that could be transferred from one material to another. Stahl used his theory to explain the following reactions, which were all of great economic importance:

- **Calcination** A metal ore heated with charcoal turns into a metal because Phlogiston is transferred from the charcoal to the ore.

- **Smelting** When a metal is heated in the air it becomes a powder because it loses its Phlogiston.

- **Making alkali** When limestone is heated to high temperatures, it changes into quicklime, because it has picked up Phlogiston from the fire.

You can investigate the theory by looking at the results of these experiments.

About the chemists

Joseph Black

(Reproduced courtesy of the Library & Information Centre, Royal Society of Chemistry.)

Joseph Black (1728–1799) was a doctor who later became an unsalaried professor of chemistry at the University of Edinburgh. His students paid a fee but his main income was from medicine. He was interested in the burning of limestone and believed in the Phlogiston theory.

Joseph Priestley

(Reproductions courtesy of the Library & Information Centre, Royal Society of Chemistry.)

Joseph Priestley (1733–1804) was an orphan from Yorkshire, who became a clergyman. Priestley enjoyed carrying out experiments and discovered that there are different types of 'air'.

Priestley publicly sympathised with the French Revolution. This caused him to be driven from his house and his library was burnt.

He went into exile in America, where with the help of his friend Benjamin Franklin, he settled. He believed in the Phlogiston theory until he died. The cartoon shows Joseph Priestley, 'Doctor Phlogiston', explaining away the Bible and other views.

Antoine Lavoisier with his wife, Marie.

(Reproduced courtesy of the Library & Information Centre, Royal Society of Chemistry.)

Antoine Lavoisier (1743–1794) was a Frenchman who studied law, but became interested in science, initially geology. He became a tax collector so that he could pay for his scientific experiments. He got married when he was 28 to a girl of 14, who became his laboratory partner. He put forward a new theory of combustion. Tax collectors were not popular in 1794 when he was guillotined.

Experiment A – Calcination (burning metals)

Work in groups, and report back to the whole class:

1. When magnesium is heated in air it will burn. Using the Phlogiston theory predict what will happen to:

    (a) the mass of magnesium
    (b) the appearance of magnesium?

2.   Carry out the experiment as described below.

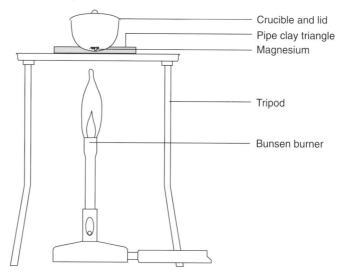

Crucible and lid
Pipe clay triangle
Magnesium

Tripod

Bunsen burner

Wear eye protection

- Clean a 10–20 cm length of magnesium ribbon with emery cloth to remove the oxide layer. Loosely coil it.

- Weigh a clean crucible and lid. Place the magnesium inside and reweigh.

- Heat the crucible for 5–10 minutes, lifting the lid a little from time to time with tongs. Ensure that as little product as possible escapes.

- Continue heating until glowing ceases.

- Cool the crucible and reweigh.

3. Carefully record your results in the table.
   Mass of crucible + lid =   g
   Mass of crucible+lid+sample =   g

	Appearance of sample	Mass of sample / g
Start		
End		
	Mass change/g	

4. Go back to question 1. Are your results the same as your predictions?

5. Do your results support or reject the Phlogiston theory?

6. Do you think you need more evidence to reach a firm conclusion? Suggest how you could get more data.

7. How could you make sure all the Phlogiston has been lost?

Experiment B – Smelting

Heating a metal ore with charcoal or natural gas

1. When copper(II) oxide is mixed with methane and heated, it will burn. Using the Phlogiston theory predict what will happen.

2. Your teacher will demonstrate the experiment.

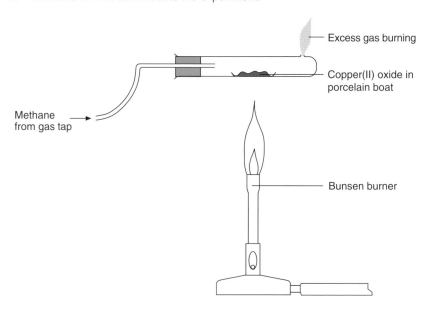

Excess gas burning

Copper(II) oxide in porcelain boat

Bunsen burner

Methane from gas tap

3. Record your observations

	Appearance of sample	Mass of sample / g
Start		
End		
	Mass change/g	

4. What can you deduce from your observations?

5. Do your results support or reject the Phlogiston theory?

6. Do you think you need more evidence to reach a firm conclusion?

7. Using the results from this experiment, what do you think Phlogiston is?

RS•C

Experiment C –
Making alkali (heating limestone)

1. Using Phlogiston theory, predict what will happen when limestone, calcium carbonate is heated?

2. Carry out the following experiment.

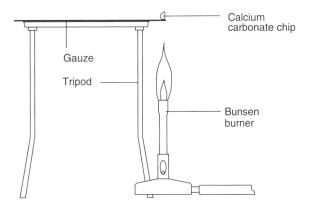

- Take 2 limestone chips, which look similar and weigh them.

- Place one of the limestone chips on the gauze as shown in the diagram and heat it over the hottest Bunsen flame for 10 minutes.

- **Do not touch the chip – it is now corrosive.**

- Let it cool down for a few minutes and then use tongs to move the chip and reweigh it.

3. Carry out the following tests on both chips.

- Compare their appearances.

- Use a nail to see if they scratch easily. Make sure you hold the chip in the tongs.

- Place the chips on a watch glass, add two drops of water to each chip and test the solution with pH paper.

Record your observations in the table.

Test	Unheated limestone chip	Heated limestone chip
Appearance		
Mass / g		Before heating: After heating:
Does it scratch easily?		
Add water & then test with Universal Indicator paper		

4. Do you think that a chemical reaction has taken place? Give a reason for your answer.

5. Do your results support or reject the Phlogiston theory?

6. Do you think you need more evidence to reach a firm conclusion?

7. Suggest how you could obtain some more data.

RS•C

Experiment C – Making alkali (heating limestone) – page 1 of 1

Burning interpretations table

Experiment	Predictions using the Phlogiston theory	Experimental results	Interpretation supports (s) or rejects (r) the Phlogiston theory
(A) Calcination	What is formed?	What is formed?	
	Change in mass?	Change in mass?	
(B) Smelting	What is formed?	What is formed?	
	Change in mass?	Change in mass?	
(C) Making alkali	What is formed?	What is formed?	
	Change in mass?	Change in mass?	

RS•C

Burning – the fact cards

Joseph Black (1)

Black worked with magnesia (magnesium carbonate) that was thought to be another form of lime. He found that when it was heated, it lost seven twelfths of its weight and changed into a new material that was insoluble in water and not alkaline. Black was puzzled because the magnesia must have absorbed a lot of phlogiston from the air and should have gone alkaline. Maybe the magnesia had lost something else?

Joseph Black (2)

Black carried out further experiments and discovered that magnesia did loose a gas when it was heated. This gas was known as 'fixed air'.

He also showed that when limestone was heated it did not absorb phlogiston but lost 'fixed air' (carbon dioxide).

Joseph Priestley (1)

Experiment

When mercury oxide was heated, the red solid decomposed and produced a colourless gas above the liquid mercury. When the gas was tested with the flame of the candle, the candle burned brightly. He had previously noted that other gases put out the flame.

Joseph Priestley (2)

Priestley found that the 'new air' would keep a mouse alive twice as long as ordinary air.

He called this 'new air' dephlogistinated air.

Antoine Lavoisier (1)

Lavoisier heard about Priestley's work and repeated it. He showed that Priestley's new gas was a component of air and that it combined with metals when they were heated. He called the new gas 'oxygen'.

Antoine Lavoisier (2)

Using sensitive scales he showed that when heated, mercury oxide lost weight as oxygen was released. He also proved that when a metal was heated in air, the metal would increase in weight by an amount corresponding to the amount of oxygen taken from the air.

Antoine Lavoisier (3)

The Law of Conservation of Matter

Matter is neither created or destroyed but is simply changed from one form into another.

Antoine Lavoisier (4)

Combustion is the combination of oxygen with other substances.

What happens when things burn? (2)

The ancient greeks thought everything was made from fire, water, air and earth.

So what happens when something burns?

The fire is released. The water and air escape. The earth or ashes are left behind.

In the 1600s, scientists thought that burning depended on air and that air was one single substance.

Robert Boyle heated tin in a sealed flask. When the seal was broken, the tin weighed more then before heating. Boyle thought the particles of fire had lodged between the particles of tin.

Georg Stahl (1660 – 1734) was a German scientist. He developed another idea. It was called the Phlogiston theory (from the Greek phlox = flame).

Every substance that burns has 2 parts – ash and PHLOGISTON.

When something burns the PHLOGISTON escapes the ash is left behind.

This gas lets things burn very brightly in it.

Something is wrong here!

Joseph Priestly worked in England. He heated mercury in air, and make a red substance. When he heated the red substance he got a new gas.

Antoine Lavoisier was a French scientist, who worked with his wife. They worked on the problem of burning and knew the phlogiston theory wasn't quite right.

Lavoisier found that the sulfur gained weight when it burnt. He thought the air was combining with the sulfur. Joseph Priestley visited Lavoisier and helped him understand.

RS•C

Read **What happens when things burn?** (2) and then answer the questions.

1. What was the Greeks theory of burning?

2. How long did their theory last?

3. What did Robert Boyle think about burning?

4. What was the Phlogiston theory?

5. What two things did Priestly find out about oxygen?

6. What was Lavoisier's new theory?

7. Why do you think Lavoisier rejected the Phlogiston theory?

Testing the Phlogiston theory

Thinking skills

Read through the information sheet **What happens when things burn? (2)**.

1. Sometimes scientists let their fondness for their theory affect their thinking. What are the faults in the Phlogiston theory explanations for the 3 experiments given at the bottom of the sheet?

2. The theory says that when a metal burns, it loses its Phlogiston. What would you expect to happen to the mass? (Increase/Decrease/No change)

Predictions for the three experiments

Calcination
Magnesium (a silvery metal ribbon) burns when heated in air.

Smelting
Copper ore (a black powder) changes when heated with charcoal or natural gas

Making alkali
Limestone (a white rock) changes into quicklime when heated in air.

Use the Phlogiston theory to make the following prediction:

Expt.	What will be formed? (Metal/alkali/powder)	What happens to the Phlogiston?	Change in mass? (Increase/decrease/ no change)
A			
B			
C			

Experiments

■ You will do or see experiment A

■ Your teacher will demonstrate experiment B

■ You will do experiment C

(A) Calcination (burning metals)
3. How would you make sure that all of the Phlogiston has been taken from the magnesium?

(B) Smelting (heating a metal ore with charcoal or natural gas)
4. What would you do to obtain more reliable results for the 'change in mass'?

(C) Making alkali (heating limestone)

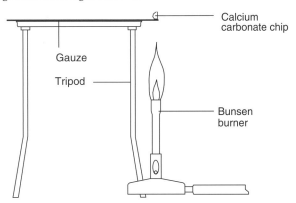

- Place a limestone chip on the gauze as shown in the diagram and heat it over the hottest Bunsen flame for 10 minutes.

- **Do not touch the chip – it is now corrosive.**

- Let it cool down for a few minutes and then use tongs to move the chip onto the watch glass.

- Add drops of water to the chip and test the solution with pH paper.

5. Do you think that a chemical reaction has taken place? Give a reason for your answer.

Results

Expt.	What is formed? (Metal/alkali/powder)	Change in mass? (Increase/decrease/no change)
A		
B		
C		Decrease

Interpretations

- Use your Predictions table and Results table to fill in the Burning: Interpretations table. In groups, discuss whether each Prediction and Result either 'supports' or 'rejects' the Phlogiston theory.

- A scientific theory is only accepted if all of the evidence fits the theory. If you decide not to accept the theory, you will need to look at more evidence and think of a better theory.

- Ask your teacher for **Burning – the fact cards** which contain further evidence. Read the information and try to think of a better theory of burning. Make sure that you can clearly explain your new theory.

Burning interpretations table

Experiment	Predictions using the Phlogiston theory	Experimental results	Interpretation supports (s) or rejects (r) the Phlogiston theory
(A) Calcination	What is formed?	What is formed?	
	Change in mass?	Change in mass?	
(B) Smelting	What is formed?	What is formed?	
	Change in mass?	Change in mass?	
(C) Making alkali	What is formed?	What is formed?	
	Change in mass?	Change in mass?	

RS•C

Burning – the fact cards

Joseph Black (1)

Black worked with magnesia (magnesium carbonate) that was thought to be another form of lime. He found that when it was heated, it lost seven twelfths of its weight and changed into a new material that was insoluble in water and not alkaline. Black was puzzled because the magnesia must have absorbed a lot of phlogiston from the air and should not have gone alkali. Maybe the magnesia had lost something else?

Joseph Black (2)

Black carried out further experiments and discovered that magnesia did loose a gas when it was heated. This gas was known as 'fixed air'.

He also showed that when limestone was heated it did not absorb phlogiston but lost 'fixed air' (carbon dioxide).

Joseph Priestley (1)

Experiment

When mercury oxide was heated, the red solid decomposed and produced a colourless gas above the liquid mercury. When the gas was tested with the flame of the candle, the candle burned brightly. He had previously noted that other gases put out the flame.

Joseph Priestley (2)

Priestley found that the 'new air' would keep a mouse alive twice as long as ordinary air.

He called this 'new air' dephlogistinated air.

Antoine Lavoisier (1)

Lavoisier heard about Priestley's work and repeated it. He showed that Priestley's new gas was a component of air and that it combined with metals when they were heated. He called the new gas 'oxygen'.

Antoine Lavoisier (2)

Using sensitive scales he showed that when heated, mercury oxide lost weight as oxygen was released. He also proved that when a metal was heated in air, the metal would increase in weight by an amount corresponding to the amount of oxygen taken from the air.

Antoine Lavoisier (3)

The Law of Conservation of Matter

Matter is neither created or destroyed but is simply changed from one form into another.

Antoine Lavoisier (4)

Combustion is the combination of oxygen with other substances.

Burning theories

Your teacher will demonstrate the first experiment and then you can carry out the second experiment with a partner.

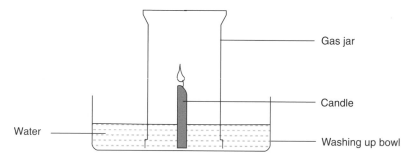

Gas jar

Candle

Water

Washing up bowl

1. On the diagram, mark in the water level at the end of the experiment.

2. The water replaces _____ gas, as the candle uses it up.

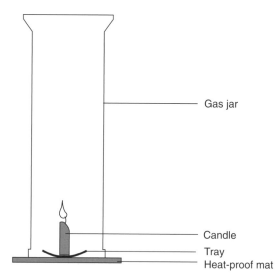

Gas jar

Candle

Tray

Heat-proof mat

■ Set up the experiment as shown in the diagram.

■ When the candle goes out, remove the gas jar and put a lid on it.

■ Add a piece of blue cobalt chloride paper. If it goes red/purple, water is present.

■ Pour some limewater into the jar and swill it around. If the limewater goes cloudy, carbon dioxide is present.

3. When a candle burns it produces _____ and

_____ gas. It gives out energy in the form of

_____ and _____.

4. Complete the word equation

   Fuel + _____ → _____ + _____

RS•C

The plastic that came out of thin air!

Teachers' notes

Objectives
■ To apply knowledge of polymerisation to other situations.

■ To learn about the discovery of Teflon®.

Outline
The story of Roy Plunkett and the discovery of Teflon® (polytetrafluoroethene) can be used to show that many things in science were discovered accidentally. This activity illustrates the way in which scientific work may be affected by the context in which it takes place. It is an example of where war influenced the development of a new material, which now has numerous applications.

Teaching topics

This activity is suitable for 14–16 year old students and could be included when teaching about addition polymerisation and the use of polymers.

Look at the student sheets before reading the more detailed teachers' notes.

Background information

Many scientists would have discarded the new gas cylinder, when no gas came out of it, but Roy Plunkett was puzzled and he wanted to find out more. He was convinced that the cylinder was full and he had to prove this to himself. If he had thrown away the cylinder, Teflon® may have not been discovered. Like Plunkett, many scientists are very inquisitive and need to understand what is going on. They will devise experiments and carry out tests until they are satisfied with the result. Sometimes they get carried away and the methods are not always that sensible. Nevertheless, Plunkett, identified a problem with the cylinder and investigated, and in the process discovered a new plastic.

Teflon® is made by DuPont.

Uses of Teflon®
Teflon® is used as a plastic coating. Probably one of the most well known examples of Teflon® coating is that of cooking utensils.

Teflon® is also used to coat many types of fabrics to meet the needs of water repellence and stain protection, without affecting the material's colour or appearance or feel in anyway. Examples are:

■ Leisure and sports wear; jackets, mountaineering clothing, suits, ties, raincoats;

■ Furnishing fabrics and leather for upholstery, curtains, wall coverings, garden furniture, mattresses, table and bed linens; and

■ Overall and work clothes treated with Teflon® protects the workers not only against weather, but also against dirt and pollution, grease, oils, chemical and acid spills.

RS•C

Teflon® has many medical applications:

- Pacemakers

- Artificial corneas

- Substitute bones for chin, nose, skull, hip

- Ear parts

- Trachea replacements

- Heart valves

- Dentures

Space applications:

- Outer skin of space suits

- Fuel tanks for space vehicles

- Heat shields on space ships

- Insulating materials for electrical cables.

The uses of Teflon® are closely related to its properties *ie* it is inert, temperature resistant *etc*. It was because of its chemical inertness that it was used in the atomic bomb for valves and gaskets. If it had not been for this, Teflon® would probably not have been manufactured on a commercial scale because it was so expensive. It was only the necessities of war that allowed Teflon® to be manufactured and developed.

Environmental impact

Polytetrafluoroethene (Teflon®) contains carbon and fluorine only, unlike chlorofluorocarbons (CFCs) where it is the presence of the chlorine atom that allows reaction with ozone (O_3) and has the effect on the ozone layer. When disposed of in landfill or in an incineration plant, fabrics treated with Teflon® decompose slowly and make a negligible difference to the total environmental impact.

Teaching tips

This activity can be used as a stand-alone activity or it can be used as homework after a short introduction to put it into context. It is intended that this activity will be used as a follow-up to addition polymerisation of alkenes or revision. It does not give enough information to answer all the questions, without some previous knowledge of unsaturated molecules, double bonds and polymerisation. However, it can be used to promote some useful discussion of the concepts covered as well as the wider issue of factors that influence the development of science.

Resources

- Student worksheets
  – The plastic that came out of thin air!

Timing

40 minutes

Opportunities for key skills

Information handling – finding out what Telfon® is used for today.

RS•C

Answers

1. 6/4/1938

2. He was developing refrigerants.

3. **a)** Tetrafluoroethene
    b) C_2F_4
    c) Unsaturated

4. The cylinder was heavy, implying it was full. The needle on the pressure gauge pointed to the full position.

5. The valve was broken.

6. The weight of the cylinder still implied that it was full. The valve was not broken and so the pressure reading was correct. Therefore there must be something inside the cylinder. He was curious.

7. He cut the cylinder in half. The cylinder could have exploded because it should have contained gases under pressure.

8. A white solid.

9. Teflon$^{®}$

10. The small gaseous tetrafluoroethene molecules much have reacted together to make a large polymer. The combination of the gas being kept under pressure and the surface of the metal cylinder to act as a catalyst may have forced the reaction.

11. Polymerisation

12. There are various things he could have done, such as heat C_2F_4 in the presence of different metals or pressurised the gas in the presence of different metals. He would have tried different conditions until he got a good yield.

13. Standard physical and chemical tests would have been carried out, such as solubility in different solvents, resistant to heat and different chemicals, strength tests *etc*.

14. See Teachers' notes.

The plastic that came out of thin air!

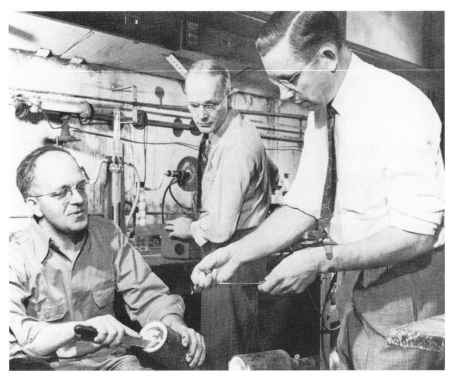

Re-enactment of the 1938 discovery of Teflon®
Left to right, Jack Rebok, Robert McHarness and Roy Plunkett
(Reproduced courtesy of the Hagley Museum and Library.)

World War II puts Teflon® on the world market

In 1938, Teflon® with its unreactive properties was very expensive to produce and was in danger of being forgotten. But General Leslie R. Grooves of the US Army heard about Teflon® and commanded it to be manufactured whatever the cost. Teflon® was used to make gaskets and valves to hold corrosive uranium compounds in the atom bomb.

It wasn't until 1960 that the public was introduced to non-stick frying pans!

RS•C

Roy Plunkett accidentally discovers Teflon

Read the cartoon and then answer the following questions.

1. What was the date?

2. What project was Roy Plunkett working on?

3. (a) Name the gas he was going to use.
   (b) Write down its formula.
   (c) Is this molecule saturated or unsaturated?

4. Give two reasons why Plunkett was surprised when no gas came out of the cylinder?

5. What did he think the problem was?

6. Why do you think he didn't just go and get another cylinder of gas?

7. What did he do next and what risk was involved?

8. What did he find inside the cylinder?

9. What did he call the new material?

10. What did Roy Plunkett think had happened to the original gas in the cylinder?

11. The new material was a plastic. Which of the following processes are used to make plastics a) cracking b) polymerisation c) fractional distillation?

12. The next day Roy Plunkett tried to make the new material in the lab. How do you think he would have done it?

13. Suggest three tests that Roy Plunkett may have carried out on the new material in order to test its properties.

14. If you have time find out what Teflon® is used for today.

RS•C

Carbon – the element with several identities

Teachers' notes

Objectives
■ To know that carbon has three allotropes.

■ To understand that the physical properties of each allotrope are dependent on the molecular structure.

Outline
The student worksheet includes information about Sir Harry Kroto, who is one of the team who discovered the third allotrope of carbon, buckminsterfullerene, and received a Nobel Prize for his work in 1996. It also includes general information about the three allotropes.

Teaching topics

This activity is suitable for 14–16 year olds and should be included when teaching about simple molecular structures and giant structures.

Sources of information

The story of the discovery of fullerenes can be read in the following books.

■ *The Age of the Molecule*, London: The Royal Society of Chemistry, 1999.

■ *Cutting Edge Chemistry*, London: The Royal Society of Chemistry, 2000.

Teaching tips

This activity could be used in lessons to create a wall display or as a homework exercise as it contains text, pictures, diagrams and questions.

The Fullerene Gallery **http://mirror.nobel.ki.se/molecules/fulleren/meny.html** (accessed June 2001), shows eight different molecules and allows students to appreciate the structure by rotating the molecules and zooming in on different parts of the structure.

Students could try and make their own models using ball and stick kits. This should help to distinguish between the different allotropes of carbon.

Resources

■ Molecular model kits (for model kits see *eg* **http://www.molymod.com**) (accessed June 2001).

■ Access to the Internet.

■ Student worksheets
– Carbon – the element with several identities.

Adapting resources

It is hoped that the format of the student worksheet given here will be adopted and that teachers will use it to make up their own pages, tailored to meet the needs of their own students and the topic being studied.

RS•C

Opportunities for using ICT

■ Using the Internet to view molecular models of fullerenes.

Answers

1. An element that can exist in different physical forms which have different physical properties. Often confused with isotopes.

2. Diamond, graphite and fullerenes.

3. a) All contain carbon atoms and covalent bonds.
   b) Diamond and graphite have giant molecular structures whereas fullerenes form simple molecules. Diamond has a three dimensional tetrahedral structure, graphite is formed in layers with a delocalised electron, C_{60} is football shaped, with a cage structure.

4. **a)** Diamond
   b) Graphite
   c) Buckminsterfullerene
   d) Graphite

5. **a)** Strong three dimensional tetrahedral structure.
   b) Delocalised electron can move throughout the structure.
   c) Small covalent molecules will dissolve in covalently bonded solvents.
   d) The weak force between the layers is broken when it is rubbed against a surface leaving the top layer of carbon atoms behind.

RS•C

Carbon – the element with several identities

Until 1985 there were two known allotropes of carbon, diamond and graphite, and now there are three. Sir Harry Kroto is one of the team who discovered the third allotrope, the fullerenes, and received a Nobel Prize for his work in 1996.

Sir Harry Kroto
(Reproduced courtesy of the Nobel Foundation)

Harry was born in Cambridgeshire. His father (a Jew) and mother fled from Berlin in the late 1930s. The family eventually settled in Bolton, where his father once again set up his balloon printing factory, in 1955. Harry worked in the factory during the school holidays.

At school, Harry enjoyed playing tennis, gymnastics and design, but he choose to take chemistry, physics and mathematics in the sixth form. The smells, bangs and flashes of chemistry attracted him to the subject, which he went on to study at Sheffield University. After gaining a first class degree, he carried on studying for a PhD, which he got in 1964. Student life was very busy, and, like all students in the 60s, playing the guitar and singing songs at parties was a must!

Harry married and his first son, Stephen was born while they were living in Ottawa in Canada, David was born when they returned to England. Harry, a humanist, is a supporter of Amnesty International and strongly believes in a secular democratic society with equal rights for all. Some words of advice from the Nobel Laureate: *"Do something which interests you or which you enjoy and do it to the best of your ability. Having chosen something worth doing – never give up and try not to let anybody down."*

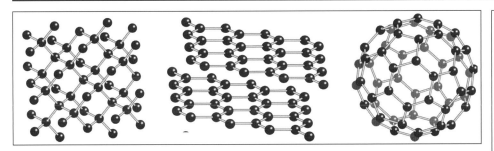

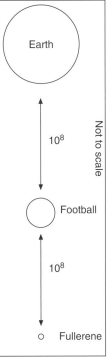

Diamond, Graphite and Buckminsterfullerene

C_{60} is the same shape as a football, but is 100 million times smaller!

Try making some carbon models.

Questions

1. What does the word allotrope mean?

2. Name the three different allotropes of carbon.

3. In what ways are the three structures a) similar b) different?

4. Which type of carbon; a) is the hardest, b) conducts electricity. c) will dissolve in oil, d) is used to write with?

5. Explain your answers to question 4.

Want to see more? Visit the Fullerene Gallery at
http://mirror.nobel.ki.se/molecules/fulleren/meny.html (accessed June 2001)

RS•C

Norbert Rillieux and the sugar industry

Teachers' notes

Objectives

- To be able to use correctly the words associated with solutions and dissolving.

- To understand the difference between a solution and a saturated solution.

- To appreciate the difficulties faced by black scientists in the 19th century.

Outline

This materially is divided into two distinct pieces of work

- **Sparkling white crystals of sugar** The student worksheet includes information about Norbert Rillieux and explains how difficult it was for him, as an African-American, to be a scientist in the 1800s.

- **Extracting sugar from sugar beet** Practical and paper based exercises on the laboratory and industrial processes.

Teaching topics

These activities are suitable for 11–14 year old students and could be included when teaching about solutions and the separation of mixtures or using neutralisation to remove impurities, using indicators to monitor the pH. It is a real application of neutralisation.

Extracting sugar from sugar beet could also be used with 14–16 year old students when teaching about rates of reaction, or acids and alkali. The extraction rate is dependent on surface area of sugar beet and this could be related to particle theory.

This activity can be used to develop practical skills such as extraction, purification, evaporation. It also presents the opportunity to show how social backgrounds can be responsible for the acceptance or rejection of a scientific idea and could be linked to citizenship.

Look at the student worksheets before reading these detailed teachers' notes.

Background information

Norbert Rillieux invents the vacuum evaporator

Rillieux's invention revolutionised the processing of sugar. In 1846 he received a patent for a multiple-effect vacuum evaporator that turned sugar cane juice into a fine grade of white sugar crystals. Rillieux's process was more efficient and economical than any other method. His basic process is still used throughout the sugar industry today.

Most of the well known scientists, engineers and inventors of the 18th, 19th and early 20th centuries are white and male, and many of them come from Western Europe. Have you ever wondered why there are not many well-known black scientists and engineers? Perhaps the answer lies in the social status of black people, especially in the

RS•C

United States of America during that time. Before the American Civil War, most black people were slaves and slaves were not considered to be people capable of being creative and having the ability to invent worthwhile things. However, all the time black craftsmen were inventing tools to help them in their daily jobs. Other people, outside of the slave communities, would not have heard about their inventions, and if they had, the inventions would not have been taken seriously. Worthwhile ideas perfected by blacks were often lost forever because of the attitude of the Federal government in the USA at the time. In 1858, Jeremiah S. Black, Attorney General of the United States, had ruled that since a patent was a contract between the government and the inventor, and since a slave was not considered a United States citizen, he could neither make a contract with the government nor assign his invention to his master. Thus it has been impossible to prove the contributions of many unnamed slaves whose creative skill has added to the industrial growth of the USA.

The national ban on patents for slaves did not apply to patents made by 'Free Persons of Colour', (or 'free Blacks') and so James Forten (1776–1842) perfected a new device for handling sails and Norbert Rillieux had no trouble in getting a patent. Rillieux's mother was a slave but his father was the slave owner and unlike most 'mixed race', he was sent to L'École Centrale in Paris to be educated. Before the end of the Civil War many of the 'free Blacks' worked to save those others in bondage. They did so by developing their literary and speaking ability rather than becoming scientists and so what they did may have limited the progress of science.

Following the Civil War, growing American industries were extensively using ideas from both black and white people and by 1913, it is estimated that black people had patented about one thousand inventions. The race of the inventor was no longer recorded on the document after Henry Blair, the first black American to receive a patent in 1834, received his second patent in 1836. Even so, it was common for inventors to take out patents in the name of a lawyer because they felt their racial identity would lower the value of the patented invention.

The following things were all invented and patented by black Americans[2]

Inventor	Invention	Date	U.S. Patent No.
E. McCoy	Lubricator for steam engines	2/7/1872	129,843
G.T. Sampson	Clothes drier	7/6/1892	476,416
R.A. Butler	Train alarm	15/6/1897	584,540
J.A. Burr	Lawn mower	9/5/1899	594,059
G.F. Grant	Golf tee	12/12/1899	638,920

And there are many other examples.

RS•C

Box 1 shows the original Rillieux evaporator.

Box 1

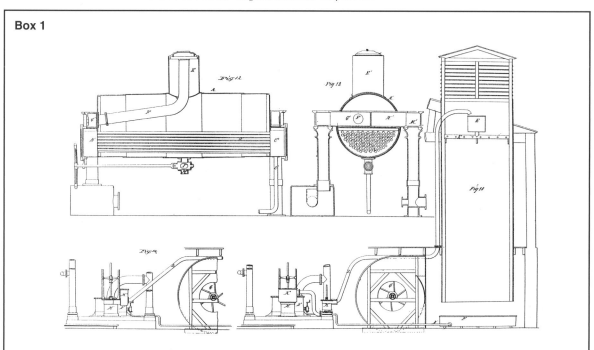

Figure 1 The Rillieux Evaporator (U.S. Patent number 4879, December 10th 1846)

A series of vacuum pans, or partial vacuum pans, have been joined together. The sugar juice in the first pan is heated and water is evaporated off. The water vapour is passed to the second where it is used to heat up the next lot of sugar juice. More water is evaporated off and once again the vapour is passed to the third pan where it is used to heat up more sugar juice. The third and last pan is connected to a condenser, and so the water vapour cools down and condenses back to a liquid. The technique works because the second pan is at a lower pressure than the first, and the third pan is at a lower pressure than the second. This means that the vapour will be 'pulled' towards the next pan. In theory there is no limit to the number of pans in the system, but in practice the number is limited because the temperature of each pan is lower.

Advantages of the system

■ **Safer** – previously slaves transferred the boiling juice from the steaming open kettle to the next by means of a long ladle.

■ **Lower boiling temperatures** resulted in greatly reduced losses of sugar in the process.

■ **Reduced costs** of labour and fuel.

RS•C

The modern process

See photocopiable sheet **The modern process of sugar extraction**.

British Sugar York Process Flows

Clean beet

Sugar beet is passed to the factory via an underground passage and "overhead trough" which cleans the beet to a certain extent (removing loose soil and trash). The beet is dried over a vibrating screen before it enters the factory.

Cossettes

After slicing, the beet takes on the appearance of thin french fries. These are called cossettes (derived from a French word) which are the optimum shape to allow maximum extraction of sugar whilst retaining mechanical stability, which is important for later pressing. At York, cossettes pass over a weigher, which gives a figure upon which to base most process flow numbers. At York, a typical flow of cossettes would be 390 tonnes per hour. This figure will be used for all following flow rates.

Diffusion

This is a process where cossettes and water pass one another in a counter-current flow. The flow of water is ratioed to the flow of cossettes using a figure called Draft. A draft of 117, which is typical for York, means that for every 100 tonnes of cassettes entering the diffuser, 117 tonnes of water will be added at the other end.

The water entering the diffuser is a mixture of borehole water (*ie* fresh water) and water returned from the pulp presses. This is called Diffusion Supply Water or DSW. DSW is typically in the pH range 5.0 to 5.7. This is controlled through the addition of a dilute solution of sulfuric acid to lower the pH. Diffusion tends to be carried out under acidic conditions to assist extraction of sugar and maintain pressability of the pulp leaving the diffuser.

Raw juice

This is the juice containing all species extracted from the cassettes. Flow of raw juice is in the region of 121% of beet sliced, which equates to 472 tonnes per hour. The pH of raw juice is in the region of 7.0 to 9.0. This is not really controlled as such, but some process streams are returned to raw juice, which can be used to alter the pH if it becomes necessary. No specific chemical is used to alter the pH.

Carbonatation

This is the primary purification stage in the process. Raw juice is mixed with a slurry of slaked lime and carbon dioxide gas is bubbled through this mixture to eliminate most of the impurities present. The slaked lime slurry is called Milk Of Lime or MOL and is made by mixing burnt lime (from the limekiln) with sweet water (*ie* water containing small amounts of sugar, to improve the solubility of the calcium). The flow of MOL is ratioed to the raw juice flow at approximately 1:10 MOL to raw juice, to a rate of around 45 cubic metres per hour of MOL.

Carbonatation is split into two stages, 1st and 2nd Carb. In 1st Carb, MOL is mixed with raw juice in a tank, which will raise the pH to around 13. This mixture is then "gassed down" with carbon dioxide gas (from the limekiln) to a target pH of 11.5. The target pH is controlled by the rate of gas addition rather than increasing or decreasing MOL flow. Residence time in the 1st Carb stage is around 10 minutes, such that a unit of juice entering will pass between mixing and reaction tanks for 10 minutes before passing forward. The resultant mixture of lime and juice is then allowed to settle in a clarifier before the supernatant liquor goes to 2nd Carb.

RS•C

In 2nd Carb, further addition of carbon dioxide gas takes place to further lower the pH and precipitate out any remaining dissolved calcium salts. Target pH in 2nd carb is 9.2, ranging from 8.5 to 9.7 dependent upon gas flow.

Thin juice

Juice leaving the 2nd Carb stage is filtered and sulfur dioxide (in a gaseous form from a sulfur stove) is added to prevent colour formation. The addition of sulfur dioxide has a tendency to depress pH, so some addition of either sodium carbonate or magnesium oxide is made to keep the pH in the range 8.0 to 9.2. The resultant juice is called thin juice, with a flow rate of around 117% beet sliced or 456 tonnes per hour.

Thick juice

Thin juice passes through a series of evaporators, which remove the majority of the water, to concentrate the juice to a level where it can be turned into crystal sugar. If the flow of thick juice leaving the evaporators is approximately 30% beet sliced (117 tonnes per hour) we see that 339 tonnes of water is removed from thin juice. No buffering of thick juice is done through the evaporators, so pH control is only done on thin juice.

The overall flow of 117 tonnes per hour of thick juice is then split between some going forward to make sugar crystal and some sent to tanks for storage and processing during the summer. This split is biased towards crystal production, with around 45% of the overall flow going to store. Before being sent to store, the juice is buffered to pH 9.2 by adding sodium carbonate solution and cooled to around 20 °C.

Juice to pans

The flow of thick juice to the sugar end of the factory is mixed with sonic sugars returning from 2nd and 3rd boiling stages, to give Juice to Pans or JTP. It is important to maintain a pH above 9 in the sugar end of the factory, so sodium carbonate can also be added direct to JTP if necessary. JTP flow is dependent upon the flow of juice to store, but is typically in the region of 65 tonnes per hour. The juice is then charged to a white pan, boiled under vacuum and seeded to produce a mixture of sugar crystals and a mother liquor. This mixture is called massecuite.

White sugar

The white massecuite is then separated into crystal and liquor in a batch centrifuge.

The liquor is reboiled to increase the overall sugar extraction, whilst the crystal is dried and then cooled before being stored in silos. The sugar produced passes over another weigher before entering the silos, at a rate of around 37 tonnes per hour, or around 10% of beet sliced.

Molasses

After the third boiling stage in the sugar end, the massecuite produced is separated into crystal and liquor in a continuous centrifuge. The crystal separated is mixed back in with thick juice and the liquor produced is molasses. Aside from white sugar and any losses in the process, molasses constitutes the only other route for sugar leaving the factory. Molasses is produced at a rate of around 11 tonnes per hour, or about 3% of beet sliced.

Sources of information

Information, Facts about British Sugar 1999/2000, Peterborough: British Sugar, 1999.

The British Sugar website **http://www.britishsugar.co.uk** (accessed June 2001).

RS•C

Sparkling white crystals of sugar

Teaching tips

This could be used as a follow up to a lesson on solutions, or a general revision lesson.

After a general discussion about Norbert Rillieux and life for coloured people in America in the 1800s, present the class with the problem Norbert Rillieux was faced with *ie* how to get white crystals of sugar.

Students could then carry out experiments A and B with the help of a partner, to see if they can obtain white crystals of sugar. To save time, half the class could carry out experiment A while the rest does experiment B, and then share the results.

Finally using the background information provided, tell the class how Rillieux solved the problem and the fact that British Sugar still base their manufacturing process on the principles thought up by Rillieux.

Rillieux's invention allowed the evaporation to take place at a lower temperature. You can obtain white crystals by putting the residue through a centrifuge and adding a small amount of water. Impurities that make the sugar go brown during the extraction (from sugarbeet or cane) process are generally the invert (*ie* glucose, fructose, and other saccharides) produced by the hydrolysis of the sucrose molecule, and the amino-nitrogen from the beet. These compounds can react together to form highly coloured species.

The experiments should be written up, using the words included on the student worksheet. You may wish to allow time to discuss the meaning of the words before they write up the experiments.

Resources

- Sugar
- Water
- Bunsen burner
- Tripod
- Gauze
- Pipe clay triangle
- Heat proof mat
- Beaker
- Evaporating basin
- Spatula
- Glass rod
- Student worksheets:
  – Sparkling white crystals of sugar (1)
  – Sparkling white crystals of sugar (2)

Timing

1 hour + homework

RS•C

Adapting resources

Sparkling white sugar crystals has been included as an example of how adjusting the required level of linguistic skills allows access to a wider range of students.

■ **Sparkling white sugar crystals 1** has been written for more able students, with a high reading age.

■ **Sparkling white sugar crystals 2** has been written for less able students, with a lower reading age.

It is hoped that the format of the student worksheet given here will be adopted and that teachers will use it to make up their own worksheets tailored to meet the needs of their own students and the topic being studied.

Opportunities for using ICT

Using the Internet to find out more about African-American scientists.

Answers

1. If Norbert Rillieux had been a white man his ideas and inventions would have been readily accepted. He would probably have been educated at home and not been sent to Paris. He would have been able to visit other scientists and sugar plantation owners freely; and he would probably have carried out more research and published more scientific papers. He might even be famous for a number of inventions now.

   Dissolve – a physical change where particles of solute mix separately and intimately with particles of solvent.

   Soluble – a substance that will dissolve in a solvent.

   Insoluble – a substance that will not dissolve in a solvent.

   Solvent – a liquid used to dissolve things.

   Solute – a substance which dissolves in a solvent to make a solution.

   Solution – formed when solids, liquids or gases dissolve in a solvent.

   Saturated – a solution that contains as much of the dissolved substance as possible at a particular temperature.

   Evaporate – when a liquid turns into a gas (at its surface).

   Condense – when a gas turns into a liquid.

   Crystal – a regular arrangement of atoms, ions, molecules or polymers.

2. In experiment A, crystals will not form and a brown sticky substance will be left behind. One of the problems is that during the heating the sugar crystals start to break down.

   In experiment B, crystals should form, but it will take a long time for them to form, as the solution cools down slowly. In practice the seeding technique is used to induce crystallisation of the thick syrup.

RS•C

RS•C

Extracting sugar from sugar beet

Teaching tips

Introduction to the lesson The lesson could start with the teacher spending 5 minutes talking about how we live in a multi-racial society where everyone has opportunity and then explain how difficult it was to be a black scientist in America in the 19th century. Introduce Norbert Rillieux and present the class with the problem he solved.

Problem In the sugar extraction process, when the final evaporation is carried out, the sugar molecules start to decompose, leaving a brown mess! This is the problem that the sugar industry was faced with, when they were extracting sugar. A 'seeding method', was used to induce crystals but the overall process was very slow.

Experiment Extracting sugar from sugar beet. Full experimental details are given in the teachers' notes. However, the procedure is lengthy and will take about three double lessons to complete the practical work. Therefore, you may wish to only carry out one part of the experiment or to use the paper exercise **Making sugar in the school laboratory**, which will help the class focus on some aspects of experimental method. This experiment is based upon material contained in the Sugar Challenge[3].

Follow up work or homework Extracting Sugar on an Industrial Scale Students should have no difficulty carrying out this work on their own, if they have completed the other activities. The industrial process follows the same procedure as on the laboratory scale. This is a good opportunity to point out that the science learnt in schools is actually used in the real world.

An alternative start to the lesson would be to follow on from the **Sparkling white sugar crystals** sheet. By then the class will have discovered the problem the sugar industry was faced with. The class can then be presented with Rillieux's solution to the problem by showing the diagram from the original patent together with an explanation. Then explain that we are going to see if vacuum evaporation really works.

Resources

- Sugar beet (or parsnips, carrots, beetroot)
- Chopping board and knife for slicing the sugar beet (or food processor)
- 1 dm^3 measuring cylinder
- Balance
- Bunsen burner, tripod, gauze, heat proof mat
- Large beaker and conical flask $(1–2 \text{ dm}^3)$
- Porcelain tile
- *Milk of lime *ie* a suspension of freshly slaked lime
- Sampling pipette
- Carbon dioxide supply connected to a perforated rubber policeman (see Figure 2)
- Thymolphthalein and phenolphthalein or a pH meter or pH data logging equipment
- Apparatus for vacuum filtration (see Figure 3)
- Safety screen
- Safety glasses

RS•C

- Glass stirring rod

- 0–110 °C thermometer

- Apparatus for vacuum evaporation (see Figure 4)

- Jam jars (for storing the juice in) and sticky labels

- Refrigerator

- Student worksheets
  – Extracting sugar on an industrial scale
  – How to extract sugar in the school laboratory

Practical tips

To slake quicklime – heat a sample of quicklime (CaO) in a crucible for 15 minutes over a hot Bunsen flame. Test the sample by placing it in a beaker and adding a little water. If the water gets hot the lime is adequately slaked. To make the suspension continue to add water until the mixture has a consistency of emulsion paint. Use a safety screen while carrying this out.

Experimental instructions

1. Preparation of the beet

- Wash the sugar beet.

- Remove the stem of the beet.

- Weigh it and record its mass.

- Chop it up into long thin 'french fry' like pieces.- this will increase the surface area.

- Put the sugar beet pieces into a beaker of boiling water and gently simmer for about half an hour, until it is soft. (Roughly 1.2 cm^3 water / g beet) – to extract the sugar.

- Separate the beet from the sugar by decanting. The impure solution can be stored for up to a week in a refrigerator.

2. Purification of the liquid extract

- Heat the liquid extract to 80 °C.

- Add about 30 cm^3 of a suspension of freshly slaked lime ($Ca(OH)_2$) – to react with the acidic impurities. This should be made up fresh before the lesson.

- Carefully bubble CO_2 through the solution as shown in Figure 2.

- Regularly take samples with a pipette and test the pH, by dropping the sample on thymolphthalein paper or using a pH meter.

- Stop adding CO_2 when the pH reaches 11.2. The indicator will be pale blue.

- At pH 11.2, filter the extract under vacuum (see Figure 3). You may need to pre-filter through a piece of linen. Use coarse filter paper.

- Bubble CO_2 through the filtrate.

- Regularly take samples with a pipette and test the pH, by dropping the sample on phenolphthalein drops on a porcelain tile or using a pH meter.

- Stop adding CO_2 when the solution reaches pH 9 – the indicator will be light pink.

■ Filter under vacuum, using a fine filter paper. The straw coloured filtrate is called 'Thin juice' and maybe stored in a refrigerator for up to a week.

3. *Concentration of the juice by vacuum evaporation*

■ Set up the apparatus as in the diagram in Figure 4. As this takes a long time, pool the juice from several groups. It takes about 3 hours for 1 dm^3 to evaporate sufficiently at 40 °C under high vacuum. So it should be left to run over the subsequent lessons or lunchtime. Use a safety screen.

■ Add the thin juice and continue evaporation until the juice has a consistency of thick porridge. 'Thick juice'.

■ Turn off the vacuum.

4. *Crystallisation*

■ Remove the thick juice and add a little icing sugar, stirring slowly.

■ Crystals will become visible in the juice, which can then be filtered off. The crystals will be a bright brown colour.

RS•C

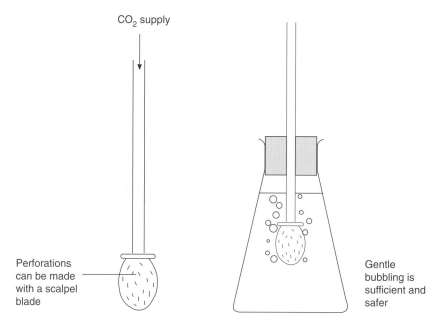

CO₂ supply

Perforations
can be made
with a scalpel
blade

Gentle
bubbling is
sufficient and
safer

Figure 2 Perforated rubber policeman for CO₂ bubbling

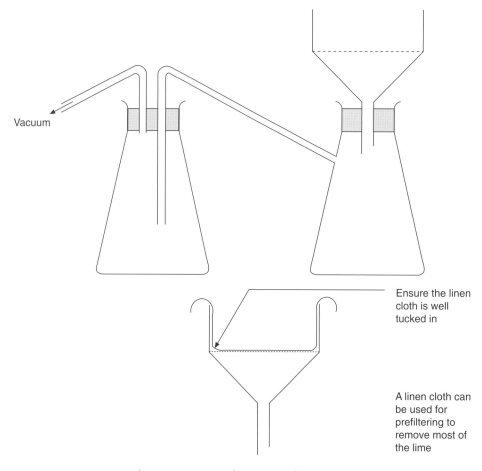

Vacuum

Ensure the linen
cloth is well
tucked in

A linen cloth can
be used for
prefiltering to
remove most of
the lime

Figure 3 Apparatus for vacuum filtration

RS•C

RS•C

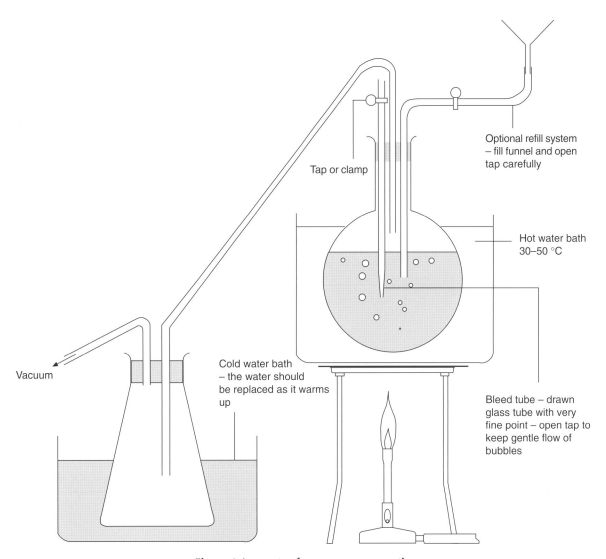

Optional refill system
– fill funnel and open
tap carefully

Tap or clamp

Hot water bath
30–50 °C

Vacuum

Cold water bath
– the water should
be replaced as it warms
up

Bleed tube – drawn
glass tube with very
fine point – open tap to
keep gentle flow of
bubbles

Figure 4 Apparatus for vacuum evaporation

Timing
The full extraction experiment will take between 6 and 8 hours.

The paper exercises **Extracting sugar on an industrial scale** and **Making sugar in the school laboratory** will take about 1 hour.

Adapting resources

Opportunities for ICT
Data logging An easy way to monitor the pH of the solution would be by a using pH probe connected to a computer. The stored data could be used at a later date to look at the rate at which the CO_2 removed the excess slaked lime.

RS•C

Answers

Making sugar in the school laboratory (a paper exercise)

1. The sugar beet is chopped up to increase the surface area during the extraction process.

2. The sugar particles dissolve in the water.

3. Lime is added to remove the acidic impurities.

4. The pH of lime is about 14.

5. Carbon dioxide is a weak acid and it is neutralised by the lime. As the concentration of lime decrease the pH also decreases.

6. Using a pH meter or a suitable indicator.

7. A vacuum lowers the boiling point of a liquid. This will allow the water to evaporate off at lower temperature so that the newly formed sugar crystals will not decompose, and not leave a brown mess in the reaction vessel.

8. Accept a description of, or a labelled diagram, showing how to filter by gravity or under vacuum.

Extracting sugar on an industrial scale

1. Sugar beet – a root crop that stores sugar in its root. The root swells up as it grows beneath the ground. It looks like a large parsnip.

   Continuous – the sugar beet extract flows directly from one place without stopping, until it reaches the end of the process.

   Crystallisation – is the process by which crystals are formed from a saturated solution.

   Batches – The reaction takes place in a big container. At the end of the reaction the product is emptied out and replaced with a new lot of reactant.

   Milk of lime – a suspension of calcium hydroxide.

   Power station – a place for generating electricity.

   Evaporators – the place where water is evaporated to leave a concentrated solution.

   National grid – the network that supplies electricity to individual houses and other buildings.

   By-products – extra products that are made during a reaction.

   Neutralise – this happens when acids and alkalis react together.

2. See diagram overleaf.

RS•C

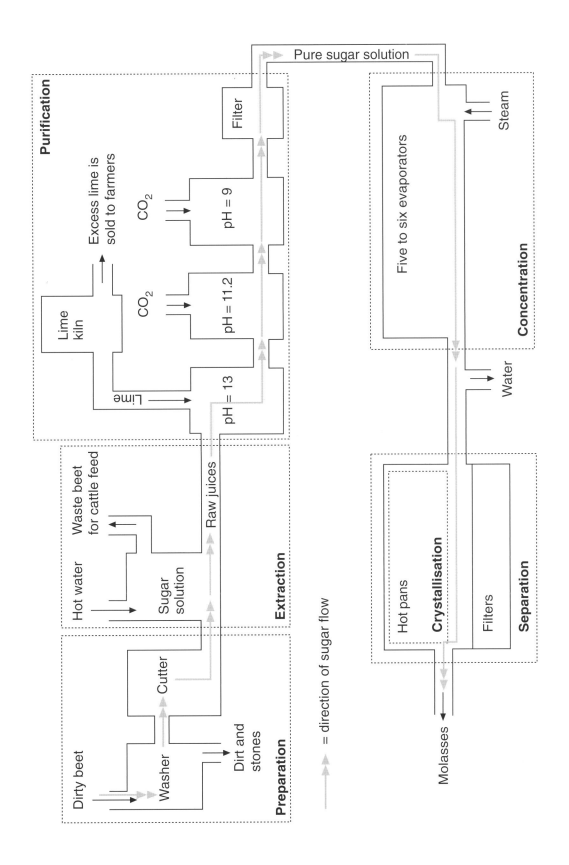

The modern process of sugar extraction – page 1 of 1

Sparkling white sugar crystals (1)

Norbert Rillieux solved the problem of obtaining white sugar crystals with his invention of the multiple vacuum evaporator in 1846. The new method was cheaper to run, safer to use and gave a better purity. Sugar factories today still use his process. Rillieux's scientific understanding was at least seventy years ahead of his time.

Norbert Rillieux
(Picture reproduced by courtesy of the Schomburg Center, New York Public Library)

Norbert Rillieux (1806–1894) Sugar Chemist and Inventor

Norbert was born in New Orleans, Louisiana. His father was an engineer and his mother a slave on his father's plantation. Norbert was not a slave but a 'free man of colour'.

He was sent to L'École Central in Paris to be educated, where at 24 he became a lecturer in applied mechanics. He published several papers on steam engines, before returning to Louisiana in 1834 and to work the sugar plantations. In 1846 he patented his vacuum evaporator.

Although recognised for his work he was socially unacceptable because of the colour of his skin. He was provided with a special house to live with servants, but he could not be entertained at the Plantation owner's house or in the house of any white man. There were also restrictions on his personal freedom, as there were on other 'free people of colour'. As social conditions got worse and the sugar industry was in decline he returned to Paris (around 1854), where he got interested in Egypt and hieroglyphics. He did not publish any more material on his sugar system until 1881 when he was nearly 75.

Norbert died in France in 1894. It has been said that Rillieux's invention of the sugar process was the greatest in the history of American chemical engineering.

Question 1. In what ways do you think that Norbert Rillieux's life would have been different if he had been a white man?

Question 2 **What do the following words mean?**	Dissolve	Solute	Condense
	Soluble	Solution	Crystal
	Insoluble	Saturated	
	Solvent	Evaporate	

Experiment A
Dissolve some sugar in water. Evaporate off most of the water. Leave in a warm place. Do crystals form?

Experiment B
Make up a saturated sugar solution. Warm up the solution so that all the sugar dissolves. Add a sugar crystal and leave to cool.

What's the problem with this method?

To find out more about African-American scientists and their social status in the 19th century visit the African Americans in the sciences website at **http://www.lib.lsu.edu/lib/chem/display/faces.html** (accessed June 2001).

Sparkling white sugar crystals (2)

Norbert Rillieux solved the problem of obtaining white sugar crystals in 1846. His method was cheaper to run, safer to use and gave a better purity. Sugar factories today still use his process. Rillieux's scientific understanding was years ahead of his time.

Norbert Rillieux
(Picture reproduced by courtesy of the Schomburg Center, New York Public Library)

Norbert Rillieux (1806–1894) Sugar Chemist and Inventor

Norbert was born in New Orleans, Louisiana. His father was an engineer and his mother a slave. Norbert was not a slave but a 'free man of colour'.

He was sent to Paris to be educated. He published several papers on steam engines. He returned to Louisiana in 1834 to work the sugar plantations. In 1846 he patented his vacuum evaporator.

He had problems because of the colour of his skin. He was provided with a special house. However, he could not be entertained in the house of any white man. As the sugar industry was in decline he returned to Paris (around 1854). He did not publish any more material on sugar until 1881 when he was nearly 75.

Norbert died in France in 1894. It has been said that his invention of the sugar process was the greatest in the history of American chemical engineering.

Question 1. How would Norbert Rillieux's life have been different if he had been a white man?

Question 2	Dissolve	Solute	Condense
What do the following words mean?	Soluble	Solution	Crystal
	Insoluble	Saturated	
	Solvent	Evaporate	

Experiment A
Dissolve some sugar in water. Evaporate off most of the water. Leave in a warm place. Do crystals form?

Experiment B
Make up a saturated sugar solution. Warm up the solution so that all the sugar dissolves. Add a sugar crystal and leave to cool.

To find out more about African-American scientists visit the website at
http://www.lib.lsu.edu/lib/chem/display/faces.html (accessed June 2001).

How to extract sugar in the school laboratory

Method

Preparation of the sugar beet

- Wash the beet and cut off the stem
- Weigh and record its mass
- Chop it up into long thin 'french fries' like pieces

Extraction of the sugar

- Put the sugar beet into a beaker of boiling water
- Gently simmer until it is soft
- Carefully pour off the sugar solution

Purification of the sugar solution

- Heat the liquid to 80 °C
- Add 30 cm^3 of freshly slaked lime (calcium hydroxide)
- Bubble carbon dioxide through the solution until the pH = 11.2
- Filter
- Bubble carbon dioxide through the solution until the pH = 9.0
- Filter under vacuum
- The straw coloured filtrate is called 'Thin juice'

Concentration by vacuum evaporation

- Set up the vacuum evaporator at 40 °C
- Add the thin juice
- Continue to evaporate until the juice has a consistency of thick porridge. This is 'Thick juice'
- Turn off the vacuum

Crystallisation

- Add a little icing sugar and stir slowly
- Wait until crystals have formed
- Filter off the sugar crystals

Questions

Preparation of the sugar beet

1. Why is the sugar beet chopped up?

Extraction of the sugar

2. What happens to the sugar particles when the beet is put in hot water?

Purification of the sugar solution

3. Why is the lime added?
4. The pH of lime is about _____

5. Why does the pH change when carbon dioxide is added?
6. How do you test for pH?

Concentration by vacuum evaporation

7. Why is a vacuum evaporator used?

Crystallisation

8. How do you filter off the crystals?

RS•C

Extracting sugar on an industrial scale

British Sugar extracts sugar from sugar beet. The method is very similar to the one you could use in the school laboratory, but obviously it is done on a much larger scale. At the sugar factory in York, 500 lorry loads of **sugar beet** arrive each day and over 9000 tonnes are processed each day, making about 800 tonnes of sugar. This means that the factory works 24 hours a day, 7 days a week during the sugar beet campaign. The process is **continuous**, until the very last stage, where **crystallisation** takes place in **batches**. The sugar beet factory is very efficient. It has its own lime kiln to produce **milk of lime** and carbon dioxide as well as its own **power station** to provide an energy source to heat the **evaporators**. Any unused electricity is fed into the **national grid**.

The **by-products** from the sugar beet are used to make a high energy animal feed, which is sold to farmers. Excess lime, from the lime-kiln, is also sold to farmers. They use it to **neutralise** acidic soil.

1. Write down the meaning of the words that are in bold in the passage.

The flow chart below shows the journey the sugar beet takes from arrival at the factory to leaving as sugar.

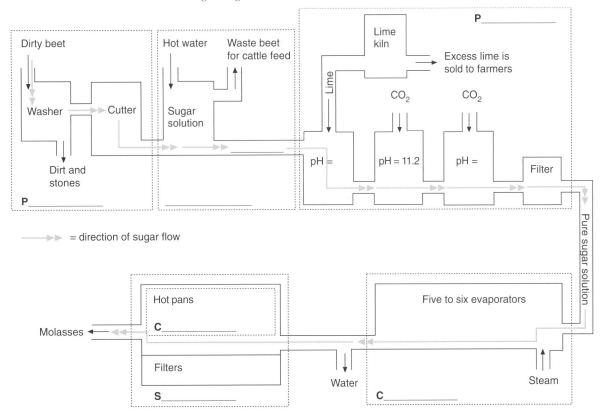

2. Use the following words to fill in the missing labels:

Purification   Preparation   Concentration   Crystallisation

Raw juices   Extraction   Separation

Mark on the chart where you think the juice will have pHs of about 13 and 9.

RS•C

References

1. M. E. Bowden, *Chemical Achievers – The Human Face of the Chemical Sciences*, Philadelphia: The Chemical Heritage Foundation, 1997.

2. I. van Sertima (Editor), *Blacks in Science: Ancient and Modern*, New Brunswick: London: Transaction Books, 1983.

3. C. Dalleywater, *Experimenting with Industry 4. Sugar Challenge*, Hatfield: The Association for Science Education, 1985. (Out of print)

4. T. Lister, *Classic Chemistry Demonstrations*, London: Royal Society of Chemistry, 1995.

5. K. Hutchings, *Classic Chemistry Experiments*, London: Royal Society of Chemistry, 2000.

RS•C